The Afterlives of Specimens

The Iowa Whitman Series
Ed Folsom, series editor

The Afterlives of Specimens

Science, Mourning, and Whitman's Civil War

Lindsay Tuggle

University of Iowa Press, *Iowa City*

University of Iowa Press, Iowa City 52242

www.uipress.uiowa.edu
Printed in the United States of America

Design by Barbara Haines

The University of Iowa Press is a member of Green Press Initiative and is committed to preserving natural resources.

Printed on acid-free paper

Library of Congress Cataloging-in-Publication Data

Names: Tuggle, Lindsay, author.
Title: The afterlives of specimens : science, mourning, and Whitman's Civil War / Lindsay Tuggle.
Description: Iowa City : University of Iowa Press, [2017] | Series: Iowa Whitman series | Includes bibliographical references and index.
Identifiers: LCCN 2017005979 | ISBN 9781609385392 (pbk) | ISBN 9781609385408 (ebk)
Subjects: LCSH: Whitman, Walt, 1819–1892—Knowledge—Anatomy. | Whitman, Walt, 1819–1892—Knowledge—Medicine. | Human body in literature. | Dead in literature. | Dead—Social aspects—United States—History—19th century. | Human anatomy—United States—History—19th century. | Literature and medicine—United States—History—19th century. | Literature and science—United States—History—19th century. | BISAC: LITERARY CRITICISM / American / General. | HISTORY / United States / Civil War Period (1850–1877).

Classification: LCC PS3242.B58 T84 2017 | DDC 811/.3—dc23
LC record available at https://lccn.loc.gov/2017005979

In memory of my sister, Amanda

CONTENTS

ACKNOWLEDGMENTS

I could write another book in gratitude to all of the individuals and institutions that made this one possible. First of all, it would not have been practically feasible without financial support from the Kluge Center at the Library of Congress, the Mütter Museum/College of Physicians of Philadelphia, the Australian Academy of the Humanities, the Eleanor Wood Bequest, and the University of Sydney.

The time that I spent in residence at the Library of Congress was utterly game-changing, and not only for access to its vast Whitman collection. I was exceptionally fortunate to be part of an extraordinary community of interdisciplinary international scholars, supported by the wonderful staff at the world's largest library. I am grateful to Alice Birney and Thomas Mann for their specialist insights into the Whitman archive, and to Mary Lou Reker for her support. Heartfelt thanks to my fellow fellows and dear friends, Stefanie Schaefer and Ilaria Andreoli, whose excellent companionship made my Kluge tenure one of the most generative times of my creative life.

Robert Hicks, director of the Mütter Museum & Historical Medical Library, believed in this project well before I had the archival evidence to support many of its claims. I have been the fortunate recipient of three Wood Institute Grants from the College of Physicians of Philadelphia, and I can honestly say that this book would be a shadow of its current form without access to their archives. Robert has been a careful reader of the drafts of this manuscript, and his medical historian's perspective was integral to its evolution. His wit, humor, and hospitality made my visits to Philadelphia a morbidly delightful experience.

Through a travel grant from the Australian Academy of the Humanities, I was able to visit the Whitman papers in the Berg Collection at the New York Public Library, the Columbia Rare Book and Manuscript Library, and the Bancroft Library (UC Berkeley). The Eleanor Wood Bequest (University of Sydney) funded my archival work at the University of Pennsylvania Rare Book and Manuscript Library and the National Museum of Health and Medicine. I am grateful to these funding bodies and to all of the librarians and archivists who helped me along the way, especially Isaac Gewirtz at the Berg Collection and Brian Spatola at the National Museum of Health and Medicine.

Finally, I am deeply indebted to the English Department at the University of Sydney for funding and hosting the doctoral research that planted the seed for this book, and especially to Kate Lilley for her insightful supervision of my dissertation.

Much of the material in chapter 2 was originally published as a journal article titled "The Afterlives of Specimens: Walt Whitman and the Army Medical Museum," in the *Walt Whitman Quarterly Review* 32 (2014): 1–35. Thanks to Stefan Schöberlein and the editorial team at the *WWQR*. Most especially, my deepest gratitude to Ed Folsom for his stewardship and faith in this project. His belief in my work astonished and humbled me and, most importantly, kept me writing. Many thanks to James McCoy and the editorial team at University of Iowa Press for their patience and wisdom in turning this manuscript into a monograph. I am grateful to Adam C. Bradford, Martin T. Buinicki, and Kenneth M. Price for their insightful and vital feedback—this is a far better book for their efforts. Special thanks again to Kenneth Price for his sound guidance and counsel when I was at a pivotal stage in writing. Thanks to Jane Goodall and Christopher Lee for inviting me to be part of the *Trauma and Public Memory* collection (Palgrave Macmillan Memory Studies, 2015). While Whitman is not the focus of that essay, many of the theories that inform this book were shaped by its writing and the excellent editorial guidance of Jane and Chris. My thanks also to Kim Roberts for her friendship and hospitality while I was living in Washington, DC, and to Martin G. Murray and the "Washington Friends of Walt Whitman" for showing me around Fredericksburg.

My work has benefited from the sound guidance of wonderful mentors who I am also graced to call friends. Since authoring his examiner's report on my dissertation, Chris Peterson's feedback has made me a better writer and a stronger scholar. In the intervening years, he has become not only a vital academic lifeline, but also a dear friend. His comments on manuscript drafts of this book were crucial to its completion. I first met Laurie Johnson at the "Mourning and Its Hospitalities" Conference in the second year of my PhD. Since then, we have contributed to two edited collections together and presented on several conference panels. His belief in my work over the years has been a sustaining force. I am grateful to Sarah Gleeson-White for her friendship and counsel since we first taught together in 2007. And to Rose Lucas, whose incisive questions sent me down an entirely new path while writing this book. Rose's insights on the third chapter were absolutely vital to its resolution. I am so thankful for her generosity and hospitality, both personally and professionally.

I am delightfully beholden to a brilliant constellation of friends and family, all of whom patiently read and reread this book and helped to shape it. Firstly, my mother, Judy Tuggle, a beautiful writer in her own right, who is always my first and last reader, and my most fiery critic. To my dear friends Adrian Jones, Radhiah Chowdhury, and Jacinta van den Berg, who were with me from beginning to end and all points in between, high and low: there are no words to adequately express my gratitude. To Kate Middleton: cheerleader, therapist, gatekeeper, spell-caster, and all around co-conspirator. To Jon Caverley, whose excellent company and counsel remains one of my greatest pleasures. To my oldest friend, James Wise, who shares all my best obsessions, and accompanied me to countless Whitmanian haunts. To my beloved sister and alter ego, Emily Helton Carr: your bravery, humor, and grace remain my touchstones. To my aunt and uncle, Marsha and Rexal Leonard, who never let me forget where I came from, and always help me to return home.

And finally, impossibly, to my husband, Michael Sloan, to whom I dedicate this book.

The Afterlives of Specimens

Specimen Interiors
An Introduction

This book emerged from the repetition of a single word, "specimen," encompassing nineteenth-century desires to collect and preserve rare, strange, or revered objects, both human and nonhuman. Writing during the bloodiest war in American history, the "specimen" allowed Walt Whitman to capture a "death-picture characteristic" of countless fallen soldiers, fusing his interest in scientific classification with an intimate form of observation.[1] The term is etymologically grounded in voyeurism. The Latin root, *specere*, means "to look or behold."[2] In Whitman's historical milieu, the "specimen" referred to a representative human, animal, plant, or mineral under observation for investigative or aesthetic aims. For Whitman this talismanic entity embodied both an "unworldly" ghostliness—the "spiritual characters" of a generation lost to war—and an intensely physical "animal purity."[3] Uniting flesh and spirit, the specimen-soldier epitomized the ideal love object Whitman conjured throughout *Leaves of Grass*—not yet a corpse, but already something beyond human, a creature all the more treasured for his ephemerality and spectrality.

Over the course of his lifetime, Whitman witnessed drastic changes in relations between the living and the dead. In the space of a few decades, anatomical dissection evolved from a punishment enacted on the bodies of executed, stolen, or unclaimed cadavers to an element of preservationist technology worthy of the presidential corpse. The extended display of Abraham Lincoln's body was made possible by recent innovations in embalming, developed on the bodies of unknown soldiers. In the intervening years, Whitman transitioned from a fervent opponent of medical body snatching to a literary celebrity who left behind instructions for his own autopsy, including the removal of his brain for scientific study. How did Whitman arrive at an understanding of the corporeal afterlife so far removed from his initial anxiety in the face of posthumous wounds? What catalyzed this startling transformation, and how did it respond to cultural changes in medical, mourning, and burial practices?[4] *The Afterlives of Specimens* establishes Whitman's role in shifting cultural understandings of the body as an object of posthumous discovery and desire.

In 1863, Whitman was living in Washington, DC, where he devoted himself to tending the scores of wounded soldiers flooding into the capital.[5] Writing to Ralph Waldo Emerson, the poet declared his intention to "write a little book out of this phase of America, her masculine young manhood . . . already brought to Hospital in her fair youth."[6] That "little book" would become a specimen herbarium of sorts, a fragmentary collection of case histories titled *Memoranda during the War* (1875–76). The Washington hospitals formed a vast convalescent network, a "City of Friends" united in suffering, "deposited here in this great, whited sepulchre of Washington."[7] In the "Calamus" cluster, published just a year before the war broke out, Whitman dreamed of an "invincible city" built on "the institution of the dear love of comrades":[8]

> I dreamed in a dream, I saw a city invincible to the
>   attacks of the whole of the rest of the earth,
> I dreamed that was the new City of Friends,
> Nothing was greater there than the quality of robust
>   love—it led the rest,
> It was seen every hour in the actions of the men of
>   that city
> And in all their looks and words.[9]

Whitman witnessed the wounded manifestation of that adhesive democracy in the intimate camaraderie that developed between soldiers in the hospitals, especially the "odd" specimens he was most compelled to document.[10] Writing to Emerson on January 17, 1863, Whitman described the hospital milieu as "the best expression of American character I have ever seen or conceived—practically here in these ranks of sick and dying young men . . . wherein this moment lie languishing . . . the imperial blood and rarest marrow of the North."[11] Portraying dying soldiers as the Union's "blood" and "marrow," Whitman renders these men as corporeal matter trapped within the national wound of fratricidal war. The medical gaze and the mourner's last look merge in Whitman's eyes to create an elegiac work that incorporates the war's "strayed dead" and their "human fragments."[12] The specimen-soldier occupies the space between science and sentiment, the historical moment of convergence when the human cadaver became both lost love object and subject of anatomical violence.

Toward the end of his life, after the poet had attained literary celebrity, Whitman reflected on his own calling toward medicine, musing that he might have been a doctor had he not become a poet: "widely opposed as science and the emotional elements are, they might be joined in the medical profession, and there would be great opportunities for developing them. Nowhere is there such a call for them."[13] Despite oppositional forces, Whitman suggests that "emotional elements" and scientific "disinterestedness" could be unified in the field of medicine.[14] Whitman appropriates surgical and curatorial rhetoric to catalog his war specimens, but he employs these tools to very different ends. *Memoranda* seeks to preserve these men not only as objects of scientific curiosity, but also as "spiritual" "beings" whose collective erasure haunted the poet for decades to come.[15]

In a letter to his mother, Louisa, Whitman illustrates the final hours of a specimen whose "strangely rapid and fatal termination" captures themes that recur throughout *Memoranda*:

> Here, now, is a specimen army hospital case: Lorenzo Strong, Co. A, 9th United States Cavalry . . . shot by a shell last Sunday; right leg amputated on the field. Sent up here Monday night, 14th. Seem'd to be doing pretty well till Wednesday noon, 16th, when he took a turn for the worse, and strangely rapid and fatal termination ensued. Though I had much to do, I staid and saw all. It was a death-picture characteristic of these soldiers' hospitals: the perfect specimen of physique—one of the most magnificent I ever saw—the convulsive spasms and working of muscles, mouth, and throat . . . . Life ebbs, runs now with the speed of a mill race; his splendid neck, as it lays all open, works still, slightly; his eyes turn back. A religious person coming in offers prayer, in subdued tones; around the foot of the bed, and in the space of the aisle, a crowd, including two or three doctors, several students, and many soldiers, has silently gathered. It is very still and warm, as the struggle goes on, and dwindles, a little more, and a little more—and then welcome oblivion, painlessness, death. A pause, the crowd drops away, a white bandage is bound around and under the jaw, the propping pillows are removed, the limpsy head falls down, the arms are softly placed by the side, all composed, all still—and the broad white sheet is thrown over everything.[16]

This passage offers a "perfect specimen" of Whitman's hospital cases, exemplifying the scenes and symptoms that unite soldiers' clinical and personal histories throughout the poet's war prose. Strong's physical perfection, marred only by the convulsions that contort his neck, echoes Whitman's descriptions of other dying soldiers as "perfect" and "noble" specimens.[17] Like many of the soldiers Whitman attended, Strong was an amputee who died due to the absence of antiseptics. Herein lies the catastrophic irony unique to the Civil War: this conflict was fought with new and lethally effective weapons such as the minié ball, yet before the discovery of sepsis and germ theory. Soldiers suffered and died in unprecedented numbers because medical discoveries on how to heal the wounded body lagged behind technological innovations on how best to kill and maim. And yet, as detailed in chapter 2, the war also ushered in an era of rapid medical progress, yielding thousands of dismembered bodies for surgical training, anonymous cadavers for experimental dissection, and pathological specimens for display in the Army Medical Museum. As chapter 3 will show, Whitman's postbellum poetics intersected with the neurological discovery of phantom limb syndrome; his hospital notebooks document the experiences of amputees suffering from limbless perceptions.

Although his deathbed is surrounded by a "crowd" of doctors, students, and soldiers rather than family, Lorenzo Strong's "painless" descent into "welcome oblivion" nevertheless adheres to the antebellum conventions of the "good death," a pivotal cultural event and literary symbol to which many aspired on their deathbeds. As Drew Gilpin Faust writes, "Dying was an art, and the tradition of *ars moriendi* [the art of dying] had provided rules of conduct for the moribund and their attendants since at least the fifteenth century."[18] Dying well entailed willingness to depart the corporeal world, remaining coherent despite pain and delirium, and demonstrating allegiance to God. The deathbed was a pivotal touchstone for mourners, who gleaned from these final words and hours clues about the decedent's past life and future afterlife.[19] As chapters 3 and 4 demonstrate, the deathbed vigil was an unbridled spectacle in antebellum society and a dominant influence on Whitman's war elegies.

Faithful Hands

Whitman inhabited a landscape of diverse mourning cultures, negotiating elaborate rituals that governed interactions between the living and the dead. Antebellum anatomists, elegists, spiritualists, and mourners shared a collective fascination with the corpse that dominated the nineteenth century. Death and burial were prominent literary tropes, not only in poetry and fiction, but also in sermons, memoirs, and biographies. The historian Ann Douglas described memoirs of the period as veritable "exercises in necrophilia" due to their erotic obsessions with the dead.[20] Mourning manuals read like etiquette books mandating ornate funeral ceremonies. Stereocards of cemeteries became increasingly popular as the death rate climbed, and the demand for posthumous photographs soared as the camera became an increasingly accessible medium. Visual images were only one of a variety of methods undertaken to retain a trace of the dead within the sensory world of the living. Jewelry containing locks of hair or images of the deceased served as tokens of ongoing devotion. The emerging faith of spiritualism sought to continue contact with the dead beyond the grave.[21]

There were obvious reasons for antebellum society's obsession with all things macabre. The average American life span had steadily declined since the late eighteenth century due to epidemics of typhoid, cholera, yellow fever, and tuberculosis, which spread rapidly through densely populated cities. The century's mortality rate exploded with the deaths of more than 752,000 soldiers during the Civil War. At roughly 2 percent of the population, in today's terms the death toll would exceed 7,500,000.[22] The scourge of disease, the devastation of an entire generation of American men, the high cost of war upon civilians, and the decimation of the landscape upon which the war was waged, all combined to cast a shroud across the century.[23]

As Adam Bradford observes, with the exception of elegies and spiritualist conjurations, antebellum mourning rituals required access to the dead:

> The bereaved needed bodies for funerals and burial, bits of hair for weavings and even painting, and clothing for producing memorial quilts and jewelry. "Traces" such as these adorned mourners' bodies and walls and functioned to make the dead a vital presence in the lives of the living . . . . Therefore, failing to witness the death and burial—along with being unable to garner any trace of the individual lost—created very real impediments to the process of mourning and had very real consequences.[24]

During the period between death and burial, the corpse retained a sacred status due to its association with the recently departed soul. The transition from the deathbed to the grave was governed by social and religious rituals for the preparation, transportation, and burial of the dead.[25] First came the "laying out" of the body: the corpse was washed and dressed in a shroud or "winding sheet," then placed in the coffin.[26] A visitation followed, while the body remained in the home for a period of time (usually between one and three days) under the watchful eyes of relatives, friends, or servants (antebellum Americans were especially fearful of live burial). This allowed mourners time to gradually accept that the soul had departed, and to psychologically prepare to relinquish the body at the gravesite. The final duty of the living toward the dead was proper interment, which, prior to the Civil War, entailed one of two outcomes: burial or entombment.[27] Most people, especially those living in rural areas, were buried below ground; enclosure in family tombs was a luxury of the wealthier class.[28] Neither cremation nor embalming was practiced prior to the war, as survivors were disturbed by the spiritual consequences of damaging the corpse.[29]

The third poem in the first edition of *Leaves of Grass* (titled "Burial Poem" in 1856) highlights the universality and frequency of death: "Not a day passes, not a minute or second, without a corpse!"[30] Whitman traces the intimate specificities of the deathbed vigil:

> The faithful hand of the living does not desert the
> hand of the dying,
> The twitching lips press lightly on the forehead
> of the dying,
> The breath ceases and the pulse of the heart
> ceases,
> The corpse stretches on the bed, and the living
> look upon it,
> It is palpable as the living are palpable.
>
> The living look upon the corpse with their eye-
> sight,
> But without eye-sight lingers a different living,
> and looks curiously on the corpse.[31]

The "faithful hand" of the attendant and his parting kiss anticipate Whitman's fidelity toward dying soldiers and foreshadow the specimen as a vehicle for retaining visual and tactile access to the dead. The posthumous gaze undergoes a haunting inversion in this poem. While the living observe the dead "with their eye-sight" another, sightless yet "palpable" spirit "looks curiously on the corpse." This dual voyeurism enacts an uncanny reciprocity: while mourners stare at the corpse, the unseen spirit looks back at them. Whitman's portrayal of the deathbed spectacle and its aftermath illuminates bereavement rituals that culminated in the "last look," offering mourners a final glimpse of their loved one before burial.[32] Whitman envisioned this scene in the opening poem of the 1855 edition of *Leaves of Grass*, imagining himself "by the coffined corpse when all is still, examining with a candle."[33]

The desire to look upon the dead often outlasted the traditional viewing period, leading to a compulsion to exhume human remains. In *Dealings with the Dead, by a Sexton of the Old School* (1856), L. M. Sargent observed the enduring melancholic attachments that compelled some mourners to seek out the decayed bodies of their loved ones. Sargent condemned the "morbid desire . . . to descend into the damp and dreary tomb—to lift the lid—and look upon the changing, softening, corrupting features . . . to gaze upon the mouldering bones."[34] Even Whitman's contemporary Ralph Waldo Emerson, the transcendentalist who believed that "material" beings were but "phantoms walking and working amid phantoms," was not immune to this macabre calling. Emerson was so distraught by the loss of his wife that he sought out her remains a year after her death.[35] His journal entry for March 29, 1832, consisted of only one line, a plaintive confession: "I visited Ellen's tomb and opened the coffin."[36] As Emerson's experience demonstrates, even the most spiritually inclined mourners were in thrall to the corpse. In his analysis of letters and diaries from the period, Laderman describes the recurring theme of mourners' "fidelity" toward the dead as "bordering on the necrophilic."[37] For vastly divergent reasons—whether to reconcile loss, for monetary gain, or to advance science—doctors, thieves, embalmers, and mourners were compelled to resurrect the dead. Whitman, with his concern for the human body in all its diversity, was certainly no exception.

A Reminiscent Memorial

The Civil War transformed relations with the dead. Once rare objects of medical theft or black market procurement, corpses now carpeted battlefields. Wounded soldiers crowded makeshift hospitals where they were often tended by inexperienced surgeons. As medicine learned from the war's mangled bodies, the bereaved public came to recognize the value of well-trained doctors and the importance of anatomy in their instruction. While resistance to experimental dissection continued throughout the century, the public also came to integrate the war's preserved specimens within existing frameworks of sentimental mourning. Visitors flocked to the new Army Medical Museum, where some expressed revulsion at the exhibition of human remains, while others recognized these military relics as a patriotic form of memento mori.

Whitman articulates the dissonance between viewing human remains as worthy of mourning on the one hand, yet subject to medical scrutiny on the other. How do we balance the psychological rupture of loss with the scientific impulse to reduce bodies to anatomical material? Whitman resists the binary division between scientific dehumanization and sentimental memorialization, casting his specimens as members of another world, outside the boundaries of temporality and corporeality. As chapter 3 argues, the Whitmanian specimen is neither human, animal, nor object; it represents an utterly different species, belonging to an "unworldly" order of "beings":[38]

> The Hospital, I do not find it, the repulsive place of sores and fevers, nor the place of querulousness, nor the bad results of morbid years which one avoids like bad s[mells] . . . . But more, a new world here I find as I would show—a world full of its separate action, play, suggestiveness—surely a medium world, advanced between our well-known practiced one of body and of mind, and one there may-be somewhere on beyond, we dream of, of the soul.[39]

Exemplifying the convergence of mourning and science, Whitman sees hospitals as navigating a borderland between the familiar, physical world of the body and the afterlife of the soul. This "medium world" recalls the rhetoric of the spiritualist movement, which promoted itself as an "experimental science," grounding its esoteric aims in "the only sure foundation for a true philosophy and a pure religion" and seeking scientific "evidence" of life after death.[40] As chapters 1 and 3 articulate, Whitman borrowed from the spiritualist language of haunting and conjuration to narrate his own dialogue with the dead. Here Whitman's convalescent specimens occupy a ghostly threshold, "advanced between" the "well-known" sphere of "body and mind" and the dreamscape that lies "beyond." The hospital transcends the designation of a "repulsive place" marked by embodied suffering; it has become in Whitman's eyes a "separate" world full of "action, play, suggestiveness." The poet takes us into a space of hiatus and trauma—a profoundly unsettling scene that is also queerly revelatory—a liminal space between the "practiced" Cartesian divide and the "new," "medium world" of the soul.

Throughout his war poetry and prose, Whitman articulates this fundamental melancholia: the possibility of pleasure, perhaps even of transcendence, through loss. In a psychoanalytic sense, we stumble into being through rupture. As subjects, we emerge from the initial wound created by probing the boundaries that separate our own bodies from those of our beloved others, confronted with the knowledge that we can exist without them—that we will one day mourn them, or they us. Recurrent experiences of grief return the subject to this subterranean yet foundational moment of loss. Each death is prismatic and infinite, recalling those that preceded it, foreshadowing those to come. And yet, in the "libidinal surge" that accompanies the first flush of mourning, we are also returned to that pre-original, unwounded state, perhaps even to what Whitman might call "oneness"—suffused with the hope that "to die is different from what any one supposed, and luckier."[41]

Reading *Leaves of Grass* alongside theoretical works by Nicholas Abraham, Judith Butler, Jacques Derrida, Sigmund Freud, Eve Kosofsky Sedgwick, and Maria Torok, I argue that Whitman's specimens inhabit the threshold between scientific exploration and melancholic attachment, embodying the intimacy of mourning in the face of anonymity and dismemberment. Whitman's attention to the experiences of soldiers dying far from home reflected contemporary anxieties about the fractured domestic rituals of grief. During the Civil War, the duties of bereavement extended beyond the deathbeds of loved ones, encompassing an ethical responsibility to the unknown soldier. In the absence of a corpse and the attendant rituals, Civil War mourners experienced the anonymous deaths and estranged burials of soldiers as a form of posthumous diaspora.

At the close of the war, when Americans began the seemingly unending task of repatriating and memorializing fallen soldiers, Whitman observed: "everywhere among these countless graves . . . in the vast trenches, the depositories of slain . . . we see, and see, and ages yet may see, on monuments and gravestones, singly or in masses, to thousands or tens of thousands, the significant word UNKNOWN."[42] In the absence of both the corpse and the grave, Whitman devised an alternative method of interment: representing countless "unfound" soldiers, the specimen body is buried within the text.[43] This duality of burial is reflective of Maria Torok and Nicholas Abraham's description of the traumatic phantom as "a memory . . . buried *without a legal burial place* . . . entombed in a fast and secure place, awaiting resurrection."[44] Whitman's "countless phantoms" are conjured by the repetitive acts of writing, reading, and rereading. In his war poetry and prose, the phantasmal aftermath of amputation signifies the erotic resonance of the body's partiality. Throughout the postbellum editions of *Leaves of Grass*, Whitman insists on the continued erotic relevance of absent bodies, ending with an avowal of the enduring posthumous presence of the author. As a poetic meditation on the psychosomatic resonance of trauma and its afterlives,

Leaves of Grass has much to teach us about the portal of loss and its infinite memorial and regenerative possibilities.

Freud initially theorized melancholia as a "pathological" displacement of the affective "work" of mourning.[45] Instead of gradually releasing libidinal attachment to the dead, the melancholic internalizes the trauma of the loved one's erasure, creating a violent lacuna within the psyche, consequentially "devouring" the ego. The binary between "successful" (temporary) grief and infinite melancholia depends on the theory that substitution of a new love object can abate traumatic rupture. Although it shares certain characteristics with mourning ("depression, loss of interest in the world . . . reduction in the sense of self"), Freudian melancholia rejects the act of cathectic replacement that allows the mourner to reenter the land of the living.[46] Simply put, the melancholic prefers the company of the dead. After his daughter Sophie died in 1920, Freud revised his thinking on posthumous attachments. Over a decade after her death, Freud wrote of his continued love for her and the impossibility of its replication:

> Although we know that after such a loss the acute state of mourning will subside, we also know we shall remain inconsolable and will never find a substitute. No matter what may fill the gap, even if it be filled completely, it nevertheless remains something else. And actually this is how it should be. It is the only way of perpetuating that love which we do not want to relinquish.[47]

As Freud came to realize, the mourner longs to retain contact with the dead at all costs. The loss of the loved one doubles as a confrontation with our own mortality and isolation. We are traumatically reminded that both our individual subjectivity and our connectivity to others rests on the precipice of fracture, imperiled by the encroaching boundaries of the temporal and the corporeal.

As chapters 1 and 3 discuss at length, the psychoanalytic terms "mourning" and "melancholia" describe a spectrum of responses to loss. If one considers the wound—the moment of traumatic rupture when the subject moves from fantasies of union with the other to experiences of dismemberment and partiality—to be necessary in the development of identity, then (as Freud finally admitted) a melancholic response to loss cannot be characterized as pathological.[48] Melancholia can thus be seen not as a destruction of the ego, but rather as an attempt to integrate the other within the psyche. Maintaining an ongoing attachment to the ghost entails a willingness to suspend linear time, to creatively enter Whitman's "medium world" of dreams.[49]

In keeping with Whitman's "unending, universal mourning," Jacques Derrida famously revised melancholic resistance as an act of revolutionary "fidelity" toward the dead.[50] Yet the "unbearable paradox" of mourning renders psychic incorporation inevitably partial, otherwise the ghost's alterity is compromised.[51] Reminiscent of Whitman's poetic representations of the mourner's last look, Derrida argues that it is not the dead themselves that we psychically incorporate, but rather a series of images. "The look that is within us is not ours . . . we look at the dead, who have been reduced to images" and they return our gaze, but we are nevertheless incapable of seeing through their eyes.[52] As Michael Nas writes, "mourning is always related to the impossible incorporation of a gaze that constitutes for us an infinite demand, a gaze that always hovers between someone and something, the completely identified and the unidentifiable, the known and the unknown."[53] Like the amputees Whitman nursed who were haunted by phantom limbs, Derridean mourning entails an impossible psychic prosthesis—the replication of a "sensory ghost" that inevitably falls short: "Ghosts: the concept of the other in the same . . . the completely other, dead, living in me."[54] Whitman returns to this paradox throughout his postbellum work. At times his "phantoms" are more real than the poet himself, though they can never reclaim the living soldiers' "animal purity."[55]

Mourning was, for Whitman, an infinite literary occupation. The poet came to understand the war as "pivotal" in the cumulative construction of *Leaves of Grass*.[56] He refigured the prewar editions as prefaces to the conflict, while the text in its entirety came to function as a poetic crypt for "hecatombs of battle-deaths."[57] Prior to the war, Whitman celebrated the symbiotic reciprocity of decomposition. As chapter 1 argues, through ecoerotic decay, corpses were resurrected as "tomb-leaves" within a landscape of desire.[58] The regenerative hospitality of burial was ruptured by unprecedented numbers of Civil War casualties. Whitman feared that the seemingly "infinite dead" would exhaust nature's capacity to incorporate their remains.[59] In the war's aftermath, Whitman shifted from a focus on cyclical regeneration ("composting") toward a poetics of preservation ("embalming").[60] Finally, as chapter 5 examines, the allure of his own "specter" captured Whitman's deathbed "fancy."[61]

Haunted by traumas witnessed in war camps and hospitals, Whitman sought, through *Leaves of Grass,* to construct a form of textual mourning in which the book replaced the body as the vehicle for absorbing the anonymous war dead. Whitman's project became an "unending" elegy, in which poems were reordered and revised in the poet's unceasing effort to (re-)create a text that could absorb so many faceless ghosts.[62] In his preface to the second annex of the "deathbed edition" (1891–92) of *Leaves of Grass*, Whitman defined his life's work as an act of mourning: "this whole book is indeed finally but a reminiscent memorial."[63] Throughout revisions and additions to *Leaves of Grass* spanning nearly forty years (1855–92), Whitman sought to preserve the dead within a collected (and re-collected) work that continually expanded in an effort to house them.

Whitman's diasporic phantoms are illuminated by Derrida's understanding of haunting as visitation: the ghost as guest, the haunted as host.[64] The anxiety produced by the ghost's arrival is heightened in the figure of the posthumous stranger, the *arrivant,* who embodies the pain of a "bottomless wound, an irreparable tragedy" that "one inherits without ever coming to terms with."[65] Through its insistent anonymity, the *arrivant* requires what Derrida terms "pre-originary" mourning, which J. Hillis Miller describes as "mourning for the death that *inhabits me*, unreachably, at every moment of my always already posthumous life."[66] Like Whitman's specimen-soldiers, Derrida's *arrivant* bears the inheritance of an infinitely open wound, one that can never be closed. Because of this perpetual ambiguity, grief for such an entity must move beyond the qualifying borders of selfhood and otherness. Such mourning would be "absolute" in the sense that it knows no "prior identity" and has no end.[67] The nameless ghost calls into question the limits of mourning as a finite process specific to a unique loss. The spectral stranger shatters the illusion that loss is a quantifiable act particular to an individual death. In the chapters that follow, I examine Whitmanian manifestations of anonymous mourning across successive editions of *Leaves of Grass.* From the "unclaim'd" corpses that littered New York City streets to the "unnamed" soldiers who filled war cemeteries, Whitman asks his readers to bear witness to the stranger's blank tomb, to mourn without the specificity of identity or location.

As Peter Coviello has argued, "virtually every strand of Whitman's utopian thought devolves upon, and is anchored by, an unwavering belief in the capacity of strangers to recognize, to desire, and to be intimate with one another."[68] Throughout *Leaves of Grass*, Whitman laments the incapacity of language to translate the spiritual and physical bonds that sometimes unite strangers: "Here is adhesiveness, it is not previously fashion'd, it is apropos; / Do you know what it is as you pass to be loved by strangers?"[69] The "adhesiveness" of stranger adoration is evident from the 1855 meditations on the "dead young men" Whitman "would have loved," to the "passing stranger" of "Calamus" (1860), the "Unknown" soldiers of *Drum-Taps* (1865), and the "phantoms of countless lost" that circulate throughout Whitman's postwar poetry.[70] Likewise, the presence of specimens endures throughout Whitman's postbellum work. He even titled his autobiography, which was largely devoted to the war, *Specimen Days* (1882). Yet there has been comparatively little critical analysis of the phenomenon of specimen collection across the Whitman canon. Mark Feldman has suggested that Whitman's specimens are objects on textual display, not unlike the cabinets of curiosities popular in nineteenth-century museums and the homes of private collectors.[71] Adam C. Bradford has compellingly argued that these collections actively resist the commodification and devastation of human bodies endemic to war: "selecting those soldier specimens he could use most productively to represent the war," Whitman creates a "psycho-textual space where he recovers value."[72] In the only book to date dedicated entirely to the analysis of death, dying, and the afterlife across successive editions of *Leaves of Grass*, Harold Aspiz argues that Whitman's conception of democracy extends beyond death, analyzing Whitman's poetic struggle to reconcile his belief in an eternal soul and in the body's divinity with the gruesome realities of death in the nineteenth century.[73] Robert Leigh Davis establishes hospital culture as central to Whitman's understanding of democracy, arguing that Whitman shifts the epistemology of romance in new, medical directions: "Whitman does not romanticize disease as a sign of creativity. . . . Nor does he pathologize homosexuality as yet another kind of affliction. Rather he shows how homosexuality, like medicine and convalescence, evokes a mode of relation based on continual risk, continual doubt."[74] M. Wynne Thomas devotes two chapters in *The Lunar Light of Whitman's Poetry* to Whitman's integration of the war into his body of work—psychically, poetically, and nationally. While traumatic memories from his hospital years repeatedly "threatened his mental equilibrium," Thomas argues that Whitman refashioned himself as a "prophet of the past," preserving his personal suffering as representative of collective memory.[75]

My thinking on postbellum nostalgia is indebted to Martin T. Buinicki's *Walt Whitman's Reconstruction: Poetry and Publishing between Memory and History*. Buinicki contrasts Whitman's derision of the "cold and bloodless electrotype plates of history" with the poet's impossible (but vital) quest to "represent the war in the far less concrete form of memory." Whitman's postwar writing expands his interrogation of "unnameable" psychic phenomena. In Buinicki's words, the poet "dwells precisely on those kinds of episodes that he repeatedly asserts can never be adequately described. Even when words fail him, he nevertheless insists on documenting the gaps."[76]

Zachary Turpin unveiled an archival discovery that cast new light on the bard's moonlighting occupation as a (pseudo)science writer. Whitman was the pseudonymous author of "Manly Health and Training" (1858), a long-forgotten health manual that "aims to teach the science of a sound and beautiful body."[77] Grounded in the era's (and the poet's) preoccupation with masculine prowess, the series covers wide and disparate territories: "a diet and exercise guide . . . an essay on male beauty, a chauvinistic screed, a sports memoir, a eugenics manifesto, a description of New York daily life, [and] a history of longevity."[78] Many of these themes recur (in vastly different language) throughout Whitman's literary prose and poetry, particularly his memoir, *Specimen Days.*

Building on the work of these scholars, this book aspires to a sustained interrogation of the specimen as material remains of the dying soldier, tracing the cultural and psychological origins of nineteenth-century America's preservation compulsion, from which Whitman's collected specimens emerged. Illuminating the influences of botanical, medical, spiritualist, and sentimental discourses on *Leaves of Grass*, I offer the first analysis of intersecting scientific and mourning communities upon the human cadaver and its abandoned parts. Elsewhere in Whitman scholarship, science and mourning have been considered independently, whereas I argue that, for Whitman, they were interdependent.[79] As symbols of embodied mourning, Whitman's specimens conjure psychic and physical attachments that were, melancholically, impossible to sever.

Histories Written on Its Leaves

In 1882 Whitman published *Specimen Days & Collect*, a volume of autobiographical vignettes that he described as "the most wayward, spontaneous, fragmentary book ever printed."[80] Comprised of notes and essays written from the Civil War onward (and incorporating *Memoranda during the War* as an entire section), *Specimen Days* is the largest and arguably the most significant work of Whitman's old age (except for the postbellum revisions to *Leaves of Grass*).[81] The book was inspired by Whitman's visit to his childhood home and family graveyards in Long Island with Richard Maurice Bucke. The structure echoes that of a scrapbook or herbarium, in which Whitman collects his own "specimen interiors":

> I publish and leave the whole gathering, first, from that eternal tendency to perpetuate and preserve which is behind all Nature, authors included; second, to symbolize two or three specimen interiors, personal and other, out of the myriads of my time, the middle range of the Nineteenth century in the New World; a strange, unloosen'd, wondrous time.[82]

Whitman articulates the "eternal tendency" that swept through his "Nineteenth century"—an unrelenting drive to "preserve" that underlies "all Nature." Ephemeral fragments are rewritten as historical artifacts—"personal" relics of a "strange, unloosen'd" time that is neither linear nor logical.[83] Echoing Walter Benjamin's notion of the nineteenth-century interior as a "a stimulus to intoxication and dream," the "lurid interiors" of Whitman's hospital prose act as a portal through which the poet hopes to salvage the war's human "debris."[84] Like Benjamin's collector, Whitman seeks to rescue these bodies from their military value as war commodities, preserving them within his own "interior phantasmagoria"—the book.[85] The Whitmanian specimen is not just an object to be looked *at* (as its etymological root suggests), but an invitation to engage in a new way of *seeing*, to inhabit a "New World" "beyond" physical or cerebral boundaries, to occupy, however fleetingly, the realm of the spirit.

Collecting natural history specimens was more than a pastime in nineteenth-century America—it was a widespread cultural obsession. Amateur and professional naturalists alike were possessed by the fervor. As Elizabeth Keeney has established, botany rapidly became the most popular recreational science of the century: "Tens of thousands of enthusiasts embraced botanizing by collecting, identifying, and preserving specimens."[86] Given Whitman's fascination with popular sciences, and his close friendship with the naturalist John Burroughs, the poet would have been familiar with the intricacies of specimen herbaria. A central aspect of botany's allure was its accessibility. It was relatively easy and inexpensive to conserve botanical specimens, especially when compared to the specialization, paraphernalia, and strong stomach required of taxidermists. After field excursions, botanical specimens were pressed and arranged into a herbarium. Flora dried between two flat, porous surfaces retained many of the characteristics of the living plant. In his *Introduction to Plant Taxonomy,* Charles Jeffrey discusses the capacity of a herbarium to reveal volumes about its subjects: "If accompanied by full field notes on place of collection, habitat, size, shape and other characters not apparent from the specimen itself, it is almost as good a source of information as the original living plant."[87] This description bears a striking resemblance to the case notes that became Whitman's *Memoranda during the War*: "I kept little note-books for impromptu jottings in pencil to refresh my memory of names and circumstances . . . . Some were scratch'd down from narratives I heard and itemized while watching, or waiting, or tending somebody amid those scenes."[88]

The queerness of Whitman's specimens (both erotic and strange) references a homoerotic culture of botanical exchange.[89] As Thomas Hallock has observed, specimens functioned as "shorthand for intimate relationships that were transacted across vast space," in which the plant acted as "the conduit of masculine, same-sex feeling."[90] This botanic reciprocity—the gifted plant prompted a letter of thanks accompanied by an enclosed specimen in return—is echoed in Whitman's epistolary relationships with soldiers and their families, and his habit of distributing gifts throughout the hospital wards.

The sexualization of botany was not limited to masculine discourses. As Elizabeth Petrino has argued, "nineteenth-century American women writers adapted floral rhetoric as a means to convey emotions and mediated passion . . . a language of gesture that implied meaning through a series of codes rather than through overt statement."[91] A lifelong gardener, at fourteen years of age Emily Dickinson was inspired by botany courses at Amherst Academy to create an exquisite floral herbarium containing more than 400 botanical specimens, which she pressed and cataloged in a green leather-bound volume. Her collection was so well preserved that the herbarium remains intact more than 150 years later, housed in Harvard's Houghton Library.

Antebellum botanists believed that each plant held a unique symbolic meaning. Dickinson's botanical textbook at Amherst, Almira H. Lincoln's *Familiar Lectures on Botany,* included a chapter on "The Symbolic Language of Flowers."[92] For Dickinson and her contemporaries, the study of botany entailed a specific nomenclature of codified floral meanings that correlated with human emotions and anatomy.[93] Dickinson equated gardeners with the "Savants" of "Comparative Anatomy." Her "meekest flowers" were "rare tenants" of wonder, no less revelatory than the "secrets" "unfolded" by articulated "bone[s]."[94] One of Dickinson's "favorite writers," John Ruskin, observed correlations between the scientific collection of plants and the botanical studies made by poets and painters. In his preface to the second edition of *Modern Painters* (1856), Ruskin depicts a collector who bestows upon his floral specimens "quasi-human" characteristics:[95]

> [He] observes every character of the plant's color and form; considering each of its attributes as an element of expression, he seizes on its lines of grace or energy, rigidity or repose; notes the feebleness or the vigor, the serenity or tremulousness of its hues; observes its local habitats, its love or fear of peculiar places, its nourishment or destruction by particular influences; he associates it in his mind with all of the features of the situations it inhabits, and the ministering agencies necessary to its support. Thenceforth the flower is to him a living creature, with its histories written on its leaves, and passions breathing in its motion.[96]

Like Ruskin's botanist, Whitman "observes every character" of his specimen soldiers. He notes their "feebleness and vigor," observes the details of their convalescent "habitats," and reports on their "nourishment or destruction" following medical treatments. His existence is consumed by the "ministering agencies necessary to support" the vast network of military hospitals. Years later, when contemplating his bloodstained hospital notebooks, the phantasmal specimen returns as a "living creature" whose histories are written, again and again, in successive editions of *Leaves of Grass.*

Whitman's "specimen cases" also reference Charles Darwin's collections from the voyage of *H.M.S. Beagle* (1831–37). Darwin's published account of the expedition, *Journal and Remarks* (1839), fused travel memoir with scientific field journal, detailing the discovery and study of natural history specimens.[97] Like the "pencillings on the spot" that informed Whitman's *Memoranda,* Darwin's field observations were recorded in small, portable notebooks, 7 by 4½ inches in size.[98] Whitman's self-fashioned notebooks were also designed to be "hurriedly written, sometimes at the clinique"; "each [was] composed of a sheet or two of paper, folded small to carry in the pocket, and fasten'd with a pin."[99] While lacking Darwin's systematic approach, Whitman employs his own descriptive methodology to catalog specimens:

> J. T. L., of Co. F., Ninth New Hampshire, lies in bed 37, Ward I. . . . Has gangrene of the feet, a pretty bad case; will surely have to loose three toes. Is a regular specimen of an old-fashion'd, rude, hearty New England country man . . . . Thomas Lindly, First Pennsylvania Cavalry, shot very badly through the foot—poor young man, he suffers horribly, has to be constantly dosed with morphone, his face ashy and glazed, bright young eyes.[100]

Whitman's notebooks include descriptions of soldiers' physical characteristics, geographic provenance, medical, military, and personal histories, and often conclude with allusions to an ethereal spirituality. This is not the prose of a detached observer. The "bright," ghostly eyes of Thomas Lindly stare back at the reader through the pages of *Memoranda*, a reciprocal last look that outlives both author and subject. Another specimen, Thomas Haley, epitomizes Whitman's capacity to extend and alter the postmortem gaze between specimen and collector, before and beyond the moment of death:

> In one of the Hospitals I find Thomas Haley, Co. M, Fourth New York Cavalry—a regular Irish boy, a fine specimen of youthful physical

> manliness—shot through the lungs—inevitably dying—came over to this country from Ireland to enlist—has not a single friend or acquaintance here—is sleeping soundly at this moment (but it is the sleep of death)—has a bullet-hole straight through the lung . . . . He lies there with his frame exposed above the waist, all naked, for coolness, a fine built man, the tan not yet bleach'd from his cheeks and neck. It is useless to talk to him, as with his sad hurt, and the stimulants they give him, and the utter strangeness of every object, face, furniture, &c., the poor fellow, even when awake, is like a frighten'd, shy animal . . . . Poor youth, so handsome, athletic, with profuse beautiful shining hair. One time as I sat looking at him while he lay asleep, he suddenly, without the least start, awaken'd, open'd his eyes, gave me a long, long steady look, turning his face very slightly to gaze easier—one long, clear silent look—a slight sigh—then turn'd back and went into his doze again. Little he knew, poor death-stricken boy, the heart of the stranger that hover'd near.[101]

Whitman's figuration of Haley as a "fine specimen" evokes a surreal dissonance—at once broadly representative and preciously specific. Peter Coviello describes this contrast as "an abstraction of particularity—*a specimen*" that "attenuates the quality of his intimacy with the dying young man."[102] Haley's fragility only serves to enhance his physical beauty, yet he is beyond the realm of language, accessible only through a network of gazes. Rendered incoherent by pain and narcotics, Haley is a "beautiful" but unreachable "animal," oblivious to the "utter strangeness" of his surroundings, including the "heart of the stranger that hover'd near." Except, that is, for one brief moment, when he abruptly wakes and fixes his devoted attendant with "one long, clear silent look." This "last look" entails a spectral duality—though delirious with "sad hurt," Haley nevertheless returns the intensity of Whitman's gaze.

Writing from the vantage of a war that was both an unprecedented medical tragedy and a catalyst for numerous scientific discoveries, Whitman articulates the pre- and postmortem experiences of soldiers through a lens and a language that is reverential, rather than diagnostic.[103] Whitman appropriates medical rhetoric, yet any sense of scientific detachment is notably absent from his hospital accounts. The poet's practice of specimen collection is irrevocably linked not only to his medical encounters, but also to his evolving understanding of the nineteenth-century homosocial experience. There is a rigorous specificity to Whitman's observations of soldiers' "physiognomy and idioms." The notebooks' painstakingly detailed (yet hurried and halting) accounts of soldier's personal biographies and medical symptoms suggest that the poet was engaged in a larger project than war documentation.

Whitman's specimens exist in a class distinct from their comrades. These case studies were designed to demonstrate a particular category of difference, influenced by Darwinian theories.[104] *The Origin of Species* was published in 1859, one year prior to the third edition of *Leaves of Grass* (1860), in which the term "specimen" first appeared.[105] Whitman had become intimately familiar with evolution by the time *Specimen Days* was published in 1882. He described Darwin and "the tenets of the evolutionists" as "unspeakably precious . . . to biology."[106] Whitman framed Darwin's significance to science in terms similar to his depiction of himself as the poetic incarnation of America: "Darwin is to me science incarnate."[107] Under the heading "The Great Unrest of Which We Are a Part," Whitman discussed the relationship between "Darwin's evolution" and "creation's incessant unrest," highlighting the cyclical and cellular correlations between "growth," "existence," and "decay," all of which he perceived to be in perpetual motion.[108] In the third edition of *Leaves of Grass*, Whitman's conception of his own infinite relationship to specimens foreshadows his tireless devotion to Civil War soldiers, as much as it recalls the continuity of evolution: "Let others finish specimens—I never finish specimens, / I shower them by exhaustless laws, as nature does, / fresh and modern continually."[109] In his mind, evolution became synonymous with progress: democratic, ecological, and literary. Speaking to Horace Traubel in 1889, the poet described the morphology of his work as an embodiment of Darwin's theory: "*Leaves of Grass* is *evolution*—evolution in its most varied, freest, largest sense."[110]

If *Memoranda during the War* and *Specimen Days* are read as documenting "the slow and successive appearance of a new species," then Darwin's theory of extinction offers further insight into Whitman's trauma surrounding the violent erasure of his specimens.[111] Whitman describes a camaraderie with soldiers near death, a binding attachment that bridged this life and the next. In this sensory landscape, amidst the "convulsive" embrace of shattered bodies, the wound opens the threshold to a "new world":[112]

> [I] find always the sick and dying soldiers forthwith begin to cling to me in a way that makes a fellow feel funny enough. These Hospitals, so different from all others—these thousands, and tens and twenties of thousands of American young men, badly wounded, all sorts of wounds . . . open a new world somehow to me, giving closer insights, new things, exploring deeper mines than any yet, showing our humanity

> (I sometimes put myself in fancy in the cot, with typhoid, or under the knife) tried by terrible, fearfulness tests, probed deepest, the living soul's, the body's tragedies.[113]

Whitman is not describing a sexual identity that is limited to genital acts, but rather the enduring union of cathectically merging with another: "I sometimes put myself in fancy in the cot, with typhoid, or under the knife." In the nineteenth century, the word "fancy" was synonymous with imagination: the creative faculty of forming images to represent "objects, events, or conditions that have not occurred in actual experience."[114] As an act of embodied mourning, Whitman seeks to "put himself" inside the wounded soldier, to endure "the body's tragedies." The war hospitals have created a "new world" where corporeal boundaries are fluid and suffering is a portal for "exploring deeper mines than any yet." Whitman's desire to cast aside his own body and adopt the pain of another recalls the invoked possession of spiritualist mediums. His habitation of the "languishing" body inverts psychoanalytic theories of incorporative mourning.[115] Rather than the mourner absorbing the lost beloved within the psyche, Whitman psychically enters the body of the dying man, where he discovers "closer insights" into "our humanity."[116]

Alongside this sense of discovery, Whitman's writings reveal his terror that, in the midst of his first glimpses at these otherworldly specimens and the "new world" they inhabited, it was already disappearing.[117] Almost an entire generation of American men faced the threat of "extinction" in the form of "lingering wounds," disfigurement, or death.[118] *The Origin of Species* described the rare but swift extinction of beings in a manner rhetorically similar to Whitman's "infinite dead":[119]

> In some cases . . . the extermination of whole groups of beings . . . has been wonderfully sudden. The whole subject of the extinction of species has been involved in the most gratuitous mystery. It is most difficult always to remember that the increase of every living being is constantly being checked by unperceived injurious agencies; and that these same unperceived agencies are amply sufficient to cause rarity, and finally extinction.[120]

Darwin's analysis of the mysterious vanishing "of whole groups of beings" is echoed in Whitman's elegiac catalogs of the hospitals' "spiritual" "beings."[121]

Death was claiming Whitman's specimens almost faster than he could document their presence.[122] From an evolutionary perspective, their erasure was catastrophic. On this point Darwin was quite clear: "species once lost do not reappear," not even if "the very same conditions of life, organic and inorganic, should recur."[123] It is fitting, then, that Whitman should devote the majority of his waking moments to the care of these men, providing whatever "nourishment" he could to prevent their deaths, and if that was not possible, helping them to die with some measure of dignity.[124] In the aftermath of the war and the closure of the hospitals, Whitman turned his attention to "the Million Dead" and their phantoms, inviting correlations between this literary haunting and the partial extinction of his specimen soldiers.[125]

Structure

As is perhaps already clear, the arguments underpinning this book cross disciplinary boundaries, uniting disparate characters and uncovering surprising connections. Given the breadth of genres, theories, and texts analyzed, my structural approach may not be immediately intuitive to readers from different fields. For these reasons—and for the benefit of those who would prefer to begin with a map in hand—the following overview should be of value.

Chapter 1 situates Whitman's early poetry and prose in response to the prevalence of medical body snatching in antebellum America, fueled by a black market that traded in illicit cadavers sourced from marginal citizens. Posthumous abandonment and medical resurrection were widely believed to have ethical, environmental, and spiritual consequences in this life and the next. Whitman's concern for peripheral bodies extended beyond the grave. While condemning the enterprise of grave robbing, Whitman's mid-century prose also expresses acute environmental anxieties arising from the decaying matter (human and nonhuman) that permeated the New York cityscape. Nineteenth-century sanitation movements arose from the fear of airborne miasmas emanating from graveyards, sewers, and other sources of decay. Whitman's optimistic understanding of evolution depends on the cyclical transformation of matter through biochemical processes of decay and regenerative growth. Appropriating contemporary scientific theories into his poetics, he believed that decaying matter was purified by absorption into the earth.[126] Conversely, Whitman understood the consequence of unincorporated decomposition to be an excessively polluted landscape. Influenced by the theory that miasma could be detoxified by flora, botanical regeneration underlies Whitman's ecoerotic anatomy in the first edition of *Leaves of Grass* (1855). Whitman utilizes the "tongues" of "grass" to digest the toxic exhalations of the dead, to literally swallow loss.

With the arrival of the Civil War, the poverty of cadavers available to anatomists suddenly became a vast surplus. Corpses littered battlefields, hospitals overflowed with casualties, and army surgeons undertook the unprecedented, large-scale collection of human specimens for medical instruction and public exhibition. Illuminating symmetries between Whitman's war writings and those of John H. Brinton, founding curator of the Army Medical Museum, chapter 2 traces the amalgam of medical and mourning rituals upon the military body and its "rejected members."[127] The two men led parallel lives during the war years, beginning with their mutual presence in the aftermath of the Battle of Fredericksburg. Given the probability that they crossed paths at Armory Square Hospital (where both were frequent visitors, once at the bedside of the same man), my findings are significant not only to Whitman studies, but also to Civil War medical and military histories. Remains from at least four soldiers that Whitman attended in the Washington hospitals became specimens in the Army Medical Museum. I examine the contrasting collections (literary and medical) that preserved the "human fragments" of these soldiers, nursed by Whitman in life and curated by Brinton after death.[128]

The unprecedented number of Civil War amputations led to an epidemic of phantom limbs, a diagnosis created by Whitman's friend and physician, Silas Weir Mitchell. Chapter 3 investigates the complexities of Whitman's underexplored relationship with Mitchell, and articulates the influences of the phantom limb phenomenon on Whitman's war poetry and prose. I analyze Mitchell's correspondence with amputees, alongside Whitman's elegies for "phantoms of countless lost" soldiers.[129] Moving beyond a necrophilic attachment to the corpse, the Whitmanian specimen parallels the embodied mourning of phantom limbs: an entity felt most acutely in absence. I read Whitman's narration of the war's "human fragments" as a discourse on the queer hospitality of "sensory ghosts" and their perpetually open wounds.[130]

In the years prior to Lincoln's assassination, embalmers claimed to have perfected the art of turning flesh to stone. Reminiscent of medical body snatchers, embalmers honed their skills on the corpses of soldiers stolen from battlefields. The popularity of the Army Medical Museum's human specimens set the stage for publications containing graphic accounts of Lincoln's autopsy and embalming. Public fascination with Lincoln's embalmed body promoted the Unionist narrative that decay of the democratic body could be prevented, that severance of the Southern states had been averted. Chapter 4 reads Whitman's war elegies for the president and fallen soldiers alongside the memorialization frenzy that swept the nation in the postbellum years. In the wake of so many unassimilated deaths and unburied remains, Lincoln's tomb, like Whitman's text, became a vehicle for incorporating the "debris of all dead soldiers."[131] The government instituted a reburial program that exhumed Union dead throughout the South, then moved them into the network of graveyards that became the national cemetery system. In *Drum-Taps* (1865) and subsequent editions of *Leaves of Grass*, Whitman constructed his own poetics of reburial, centered on the "double grave" of the unknown soldier.[132] The rearrangement of poems throughout Whitman's lifelong revision process recalls the resurrection and reburial of Union soldiers. The relocation of their literal and textual bodies exemplifies the diasporic afterlives of war specimens.

In the "deathbed edition" of *Leaves of Grass* (1891–92), Whitman not only memorialized the past, he also sought to secure his own future celebrity.[133] The poet's fixation on the elaborate construction of his own tomb parallels his farewell annex to the final edition, "Good-Bye My Fancy." Through the auto-elegiac rhetoric of "last words," Whitman seeks to preserve his own textual ghost, and to safeguard his posthumous legacy. Chapter 5 establishes the canonization of Whitman's autopsy by his literary disciple, Horace Traubel, as a climactic moment in the collision of medical and mourning cultures upon the cadaver at the close of the nineteenth century. Reverentially cataloging the removal of each organ, Traubel mourns every part of the body that inspired his devotion, recalling Whitman's empathic attention to the severed limbs of soldiers. Through the dissection of Whitman's physical body and the dissemination of his poetic body of work, Traubel frames Whitman's autopsy as the last, scientific "benediction" of a literary celebrity.[134]

1 Tomb Leaves

The Anatomy of Regeneration

Leaves of Grass begins in a cemetery. The poet reclines on the ground, "observing a spear of summer grass."[1] Contemplating the strangers buried beneath him, Whitman muses on the hospitality of decay. "Song of Myself" arises from the question: "What is the grass?"[2] The poem's sprawling answer arrives at a celebration of ecological transcendence over death. The origins of this elegiac inquiry began over a decade before the first edition of *Leaves of Grass* (1855). In January 1842 Whitman published a strange tale of dual burial in the *Democratic Review*.[3] "The Tomb Blossoms" describes a chance encounter in an "ancient field of the dead" on the outskirts of a "country village" untouched by modernity: "no mortar, no bricks, no gas—no newness!" The unnamed narrator sets out on a walk through the "pure air" to recover from a "few days visit to New York," which left him "sullen," "tired," and "out of humor."[4] He stops to rest outside the local graveyard, where he contemplates death at the cemetery gates: "O! fearful arch! if there were for thee a voice to utter what has passed beneath and near thee—if the secrets of the earthy dwelling that to thee are known could be disclosed—whose ear might listen to the appalling story and its possessor not go mad with terror?"[5]

He is startled from this reverie by a glimpse of movement "among the tombs," a widow "passing and repassing constantly between two and the same graves." Approaching the scene, he recognizes her as "a very old inmate of the poor-house, named Delaree." She has covered two adjacent, unmarked graves with a blanket of flowers. The narrator recalls the history of Delaree and her late husband, "natives of one of the West India Islands," who migrated to America "in order to gain a livelihood," only to find themselves ostracized by the local community, "until at last their fortunes became desperate." "Country people," Whitman observes, "seldom like foreigners."[6] After her husband died "a victim of poverty," the widow convalesced at the local almshouse. Assuming that she is adorning the graves of her husband and child, the narrator enquires about her "tomb-ornamenting" ritual. Yet, when asked whose graves she tends, the widow replies, "My husband's. . . . None but Gilbert's." Eventually, the mystery surrounding her grief is revealed: "When her husband's death occurred, she was herself confined to a sick bed, which she did not leave for a long while after he was buried." Once well enough to visit the cemetery, Delaree was shocked to find that her husband's grave could not be identified:

> With the careless indifference which is shown to the corpses of outcasts, poor Delaree had been thrown into a hastily dug hole, without anyone noting it or remembering which it was. Subsequently, several other paupers were buried in the same spot; and the sexton could only show two graves to the disconsolate woman, and tell her that her husband's was positively one of the twain.[7]

Throughout her illness, Delaree had "looked forward to the consolation" of visiting her husband's grave "as to a shrine." Devastated by the anonymity of his burial, she sought the "consent of the proper functionaries that the graves might be opened, and her anxieties put to rest!"[8] After her request for exhumation was denied, the widow remained devoted to her husband's shared grave. "Every Sunday, in the mild seasons, she went forth early, and gathered fresh flowers, and dressed both the graves . . . ever careful to have each tomb adorned in exactly the same manner." This ritual transcends the locality of her husband's body. Her cemetery vigils facilitate an ancestral connection: "In a strange land, among a strange race, she said it was like communion with her own people to visit that burial-ground."[9] Delaree mourns not only the loss of her husband, but also the diasporic rupture that prevents her return home. Despite her longing to know the exact location of his corpse, she speculates that God may have withheld this knowledge, "lest grief over it should become too common a luxury . . . and melt me away." In the end, her dying wish was to be incorporated within the double grave: "Her last desire, and it was complied with—was that she should be placed midway between the two graves." Not long after this meeting, Whitman tells us, Delaree joined her husband and his anonymous companion.

The widow's posthumous fidelity banishes the narrator's initial "terror" of the grave and its "secrets": "What wondrous thing is human love! . . . Here is this aged wayfarer—a woman of trials and griefs—decrepit, sore, and steeped in poverty—the most forlorn of her kind, and yet . . . the memory of her love hovers like a beautiful spirit."[10] His revulsion toward death is eclipsed by "wonder" at the continuity of mourning. This story offers an early example of Whitman's faith in the hospitality of burial: "The grave—the grave. What foolish man calls it a dreadful place? It is a kind friend, whose arms shall compass us round about, and while we lay our head upon his bosom, no care, temptation nor corroding passion shall have power to disturb us."[11] Whitman conceptualizes burial as the assimilation of the other into the surrounding landscape, eventually into one's own body: "And as to you corpse I think you are good manure, but that does not offend me, / I smell the white roses sweetscented and growing."[12] Eco-erotic decay is central to Whitman's sense of mourning as both panoramic and interior, both degenerative and transcendent. The earth invites the dead; their decomposition is rendered phantasmal. Through incorporative composting, anonymous corpses are transformed and revived. Grass becomes a regenerative memorial, "the beautiful uncut hair of graves."[13]

This chapter analyzes the ecological, erotic, linguistic, and scientific influences of burial practices on the first edition of *Leaves of Grass*, drawing correlations between the phenomenon of unutterability and the earth's absorptive capacity. I analyze Whitman's representations of the body's divinity (pre- and postmortem) alongside his curiosity in the face of new approaches to medical exploration. "The Tomb Blossoms" anticipates the cycles of decay and regeneration that recur throughout *Leaves of Grass*.[14] Almost two decades later, the widow's ritual is inverted in "Scented Herbage of My Breast" (1860). "Tomb-leaves" "emerge" from corpses, rather than being laid upon the grave by mourners:

> Tomb-leaves, body-leaves, growing up above me, above
> death,
> Perennial roots, tall leaves—O the winter shall not
> freeze you, delicate leaves,
> Every year shall you bloom again—Out from where
> you retired, you shall emerge again;
> . . . .
> Yet you are very beautiful to me, you faint-tinged
> roots—you make me think of Death,
> Death is beautiful from you—
> (what indeed is beautiful, except Death and Love?)
> O I think it is not for life I am chanting here my
> chant of lovers—I think it must be for Death,
> For how calm, how solemn it grows, to ascend to the
> atmosphere of lovers,

Death or life I am then indifferent—my Soul declines
to prefer,
I am not sure but the high Soul of lovers welcomes
death most.[15]

Delaree's "blossoms" have morphed into perennial "body-leaves," but their regenerative consolation remains unchanged. From a floral blanket woven by a devoted widow to "scented herbage" that "ascends" from graves to perfume the "atmosphere," Whitman assures us that "Death" remains as "beautiful" as "Love."

Exploring the neglected influences of body snatching on Whitman's work, this chapter traces the anatomical and sentimental origins of this elaborate mourning wreath, and studies the conditions of its resurrection as flora that "bloom again," year after year, from the mouths of the dead.

Guardians of the Grave

The grieving widow is a recurrent figure throughout Whitman's poetry and prose. Delaree's "white hairs" and "pale blossoms" presage the "white hairs of old mothers" that merge with the grass to become "the beautiful uncut hair of graves" in "Song of Myself."[16] She foreshadows the collective bereavement that engulfed postbellum America in the wake of a generation of lost soldiers: "over the whole land . . . an unending, universal mourning-wail of women, parents, orphans."[17] A few months after "The Tomb Blossoms," another vigilant cemetery widow surfaced in Whitman's journalism. On April 1, 1842, an article provocatively titled "Guardians of the Grave" appeared in the *New York Aurora.* Whitman recounted a cryptic tale of maternal fury that unfolded two nights earlier at Christie Street graveyard, where "a woman, armed with a pistol, guard[ed] the graves of her husband and children."[18] The impending desecration of this cemetery was not at the hands of body snatchers, but builders: "a chartered company" intended "to break up the ground of a large grave yard there, for the purposes of building upon the locality it occupies." They were met with a local militia intent on preventing the exhumation by any means necessary. Whitman's "thrill" at the widow's armed vigil echoes Delaree's posthumous fidelity: "A mysterious thing is woman's love! Here comes a widow, her husband dead, perhaps for years and years—and at the most distant rumor of insult offered to his shapeless and decaying ashes, the old tenderness and the old sympathies are roused again! Pale with excitement, she arms herself with deadly weapons, and stands over his grave, and the graves of her children."[19]

In nineteenth-century New York, widows were not the only ones keeping watch over the dead. Although the villains in this instance were neither anatomists nor their suppliers, Whitman's closing remarks invoke the threat of body snatching: "For there is in every man's breast a sentiment which leads him to regard with horror any desecration of the dark and ghostly grave. . . . Coarse must be the character, and callous the soul, that would touch sorely upon these hallowed sympathies."[20] This moral "horror" echoes popular anti-dissection rhetoric, which portrayed the practice as a defilement of the human body. An article in *Greenleaf's New York Journal and Patriotic Register* on August 8, 1797, anticipated Whitman's outrage: "Nothing appears more shocking to human nature so much as violence used toward the dead."[21] The phobic cultural response to anatomy as atrocity cannot be overstated. Fears of posthumous violence permeated antebellum society, alongside reverence toward the deathbed and the art of dying well.

Throughout the nineteenth century, when anatomical specimens were in high demand but were notoriously difficult to acquire, a black market trafficked in human remains. Expanding medical schools exhausted cadaver supplies, forcing anatomists to undertake drastic measures.[22] Body snatching was rampant in antebellum America. Whitman's revulsion toward the practice was captured in "The Eighteenth Presidency" (1856), a political tract that included "body snatchers" alongside "robbers, . . . murderers, . . . disunionists . . . [and] slave catchers" in a catalog of "dictators" who were "poisoning the politics of these states."[23] Framing dissection as a necrophilic exploitation of the dead, protesters often associated anatomists with other "skin trades" such as slavery and prostitution.[24] Dealers in the sale of human bodies (pre- and postmortem) were spoken of in the same breath as thieves, murderers, and—perhaps most severely in Whitman's eyes—traitors who threatened the Union.

To twenty-first-century readers, the correlation of slavery with anatomy (a science that, despite the questionable methods of obtaining corpses, ultimately aimed to alleviate human suffering) is counterintuitive. If one is traumatized by posthumous violence, how can one accept violence enacted upon living bodies? *Leaves of Grass* unsettles this flawed logic through empathic catalogs of "outcasts"—both living and dead. While Whitman was not religious, he remained concerned for the sanctity of the body before and after death, questioning "whether those who defiled the living were as bad as they / who defiled the dead?"[25] Whitman interrogates a system that deters (and eventually legislates against) the dissection of privileged cadavers, yet permits the sale of black bodies on the auction block. As the poet's catalogs demonstrate, marginal Americans were exploited in countless ways. The urban poor lived and worked in dangerous, unsanitary conditions. Children were forced into factory labor; foundlings and orphans crowded into asylums. Slavery and its institutional tortures haunted African Americans for generations. Genocide and disease decimated Native American populations. And yet mainstream society professed enduring reverence for the elite dead, while largely ignoring (or endorsing) these atrocities. As we shall see, antebellum decedents were as segregated as the living: those whose disenfranchisement rendered them commodities, and those protected by power and privilege that extended beyond the grave.

Fear of dissection fueled hostilities toward surgeons and their students well into the nineteenth century. The period between 1765 and 1855 saw at least seventeen anti-dissection riots, including the Doctors' Mob of 1788.[26] Writing under the pseudonym Velsor Brush in 1862, Whitman recounted the evolution of Broadway Hospital, the first major hospital in New York City: "there is a long history, not without romantic incidents, of the earlier years of the institution . . . including the 'Doctors' Riots,' which created so much alarm, and were so celebrated in their time."[27] Various origin stories surround this largely forgotten moment in American medical history. J. T. Headley's *The Great Riots of New York* (1873) suggests that the uprising was sparked on Sunday, April 13, when a group of boys playing outside Broadway Hospital peered into the window of an anatomist's chamber.[28] The surgeon brandished a dismembered arm to frighten them away. One child relayed the incident to his recently bereaved father, who exhumed the coffin of his late wife, and found it empty. The widower and a group of fellow masonic workers stormed the hospital, swelling in numbers as they marched. In a letter to the governor of Virginia, Colonel William Heth described the ensuing scene:

> In the Anatomy room were found three fresh bodies—one boiling in a kettle, and two others cutting up—with certain parts of the two sexes hanging up in a most brutal position. The circumstances, together with the wanton and apparent inhuman complexion of the room, exasperated the Mob beyond all bounds, to the total destruction of every anatomy in the hospital.[29]

While most of the surgeons had already fled the scene, Dr. Wright Post and three students remained behind to guard their collection of rare pathological specimens. Their efforts were in vain: the medical relics were burned in a street-side pyre. Post and his students would likely have met with the same end were it not for the intervention of the sheriff, who placed them in protective custody.[30] The next morning, an estimated 5,000 rioters stormed the jail, demanding that the surgeons be handed over to vigilante justice. The mob clashed with an armed militia, resulting in at least five deaths.[31] In the riot's aftermath, cemeteries were flooded with mourners inspecting graves for evidence of disturbance. After dark, armed gangs known as "Dead Guard Men" patrolled local burial grounds.[32]

Within a year, New York passed the Anatomy Act of 1789, which outlawed body snatching and allowed surgeons to legally dissect executed criminals. The bill's full title reflects not only widespread anatomical anxiety, but also reverence toward the grave as hallowed ground: "An Act to Prevent the Odious Practice of Digging up and Removing for the Purpose of Dissection, Dead Bodies Interred in Cemeteries or Burial Places."[33] Despite this newly legal source of cadavers, the legislation failed to suppress medical grave robbing. Execution did not produce anywhere near the number of corpses required, and scientists continued to plunder potter's fields and poorhouses in search of specimens.[34]

A unique lexicon emerged to describe the medically motivated theft of human remains. Protesters commonly referred to the practice as "body snatching." Medical exhumantionists were also deemed "night doctors," "fishermen," "grabs," and "sack-em-up men."[35] Another frequent designation was "grave robbers," which referred to the theft of human bodies *as property*, excluding personal effects. Immediate disposal of such artifacts was imperative, because their discovery enabled identification. Body snatchers had no interest in trinkets; the corpse alone was most valuable to science.[36] Seeking to elevate the rhetoric of disinterment, medical grave robbers deemed themselves "resurrectionists," a pun on Christian mythologies surrounding the end of days, which was widely believed to entail a literal rising from the grave.[37] Anatomists and their enablers were cast by protesters as usurpers of divinity, perpetrating assaults on spirit and flesh. Antebellum mourners perceived grave robbing as theft not only of the body, but also of the soul, since dissection theoretically destroyed the possibility of holy rapture.[38]

Whitman was personally acquainted with a notorious grave robber, albeit one motivated by celebrity rather than science. In 1831, at age twelve, Whitman began an apprenticeship as a journeyman printer at the *Long Island Patriot*, a weekly publication associated with the Kings County Democratic Party. During Whitman's employment (1831–32) the *Patriot* was overseen by the charismatic editor Samuel E. Clement, a Quaker of Southern ancestry who also served as Brooklyn's postmaster. Shortly after Whitman's arrival, Clement's reputation suffered a dire blow when he was branded a grave robber. The incident left a lasting impression on the young poet. More than twenty-five years later, he wrote a detailed account of the "revolting affair" for the *Brooklyn Daily Times* (1857): "Several gentlemen were very anxious to have the sculptured presentment of Elias Hicks, the renowned preacher of 'inner light,' who had then lately died at Jericho, Long Island." The resurrectionists included Clement and Henri Browere, a prominent sculptor of life masks whose subjects included Lafayette, Jefferson, Hamilton, and Dolly Madison. The men stole into the cemetery, unearthed Hick's corpse, and took plaster casts of his face: "From this mold a permanent one was made and several busts of Elias were formed, quite perfect it is said. But soon a quarrel arose, in reference to the division of the anticipated profits. . . [and] the molds and the few busts made from them were all smashed to pieces!" Whitman must have reflected on this scandal near the close of his life as he was writing his own biographical sketch of Hicks. The poet possessed a large plaster bust of the preacher, "one of my treasures," displayed in the parlor at his Mickle Street home.[39] The Clement affair foreshadows Horace Traubel's fetishization of Whitman's autopsied corpse and death mask, as I discuss in chapter 5.

The widow of Whitman's "Guardians of the Grave" recalls another cast of cemetery sentinels. Mourners often employed armed grave watchers to protect the recently deceased. The fees commanded by grave watchers were so high that only the wealthy could afford their services. Others, like Whitman's "guardian," were compelled to take matters into their own hands. The wealthy invested in defensive coffins, night watchers, and guard dogs. The poor, with fewer resources, devised creative solutions. *Freedom's Journal*, an early nineteenth-century African American newspaper, promoted a "cheap and easy way . . . to secure dead bodies." Corpses were buried deeply, interspersed with layers of straw to slow potential snatchers' progress, so that "the longest night will not afford time sufficient to empty the grave."[40] Another protective ritual involved the arrangement of an elaborate pattern of stones, shells, twigs, or flowers over the grave in order to detect disturbance, a practice virtually identical to Delaree's "tomb-ornamenting."[41] Resurrectionists would have to diagram this pattern so that it could be re-created after the body was removed.[42] Unadorned graves were preferable, and the ritual proved a reasonable deterrent.[43] Delaree's "tomb blossoms" may have served a dual purpose, not only to decorate her husband's grave, but also to prevent resurrection. For obvious reasons, medically motivated thefts occurred shortly after (or prior to) burial. By the time Delaree visited the cemetery after her own convalescence, the grave would already have been emptied if it had been targeted. The managers of almshouses, potter's fields, and morgues often collaborated with local medical schools to supply the "corpses of outcasts" in their care.[44] This widespread conspiracy offers one possible explanation for the sexton's refusal to exhume Gilbert's body: it may not have been there.

As Whitman's "Guardian of the Grave" demonstrates, many antebellum mourners exhibited a prolonged fear of exhumation, a threat that psychologically extended beyond the period immediately following burial, when the corpse was likely prey for body snatchers. How, then, did anatomists acquire cadavers without continually inciting riots? The solution to the problem of public panic was to steal the dead of marginalized communities who could offer little resistance. The bodies of African Americans, immigrants, prostitutes, Native Americans, and the poor, who crowded almshouses in life and potter's fields in death, were ideal targets.[45] A disproportionate number of immigrants fueled the illicit cadaver market. By 1880, only one-eighth of the population was foreign-born, yet immigrants (like the Delarees) made up one-third of the residents of poorhouses.[46] Their bodies were buried anonymously in potter's fields—the name historically given to cemeteries for the indigent and unknown.[47] Anonymous graveyards were a primary source of anatomical material. Country cemeteries, like the location of "The Tomb Blossoms," were also attractive targets due to their isolation. "Those in charge of morgues, the dead rooms of hospitals, and potter's fields, could tell some startling things about how bodies disappear from those places," a whistle-blowing doctor asserted in an 1879 issue of *Penn Monthly*. "The number of bodies that are allowed to go into the potter's fields throughout the country is very small, and the majority of those that reach them are not allowed to rest in them many hours."[48] In 1880, the anatomy demonstrator at the University of Michigan promised the university trustees that "better people" could rest assured. Although the legal supply of corpses fell drastically short, he sourced additional cadavers from the "paupers and friendless dead" of the "county houses and asylums."[49]

As early as 1765, doctors were defending their right to exhume dispossessed bodies, while claiming to leave those buried in private cemeteries unmolested. Dr. William Shippen Jr., one of the first American anatomy professors, became entangled in a debate surrounding anatomical material as private property versus public domain. Shippen studied medicine in Edinburgh and delivered his first anatomy lectures in Philadelphia in 1762. He advertised these lectures in the *Pennsylvania Gazette* (November 11, 1762), advocating their benefits not only for "young gentlemen now engaged in the study of Physic," but also as "entertainment [for] any gentlemen who may have the curiosity to understand the anatomy of the Human Frame." Shippen's promotion of anatomy as "entertainment" for "curiosity" seekers anticipates the fascination with human specimens that flourished in the postbellum years. The doctor's decision to open his lectures to a nonmedical audience was probably influenced by the popularity of public anatomy demonstrations in London and Edinburgh. He would soon discover, however, that the American public were even less tolerant than their British counterparts. On September 26, 1765, the *Gazette* carried another advertisement for Shippen's lectures, but this time the doctor included a disclaimer:

> It has given Dr. Shippen much Pain to hear that notwithstanding all the Caution and Care he has taken to preserve the utmost decency in opening and dissecting dead Bodies, which he has persevered in chiefly from the Motive of being useful to Mankind, some evil-minded persons . . . have reported to his Disadvantage that he has taken up some persons who were buried in the Church Burying Ground. . . . The Doctor improves this Opportunity to declare that the Report is absolutely false; and to assure them that the Bodies he dissected were persons who had willfully murdered themselves or were publicly executed, except now and then one from the Potter's Field . . . and that he never had one Body from the Church or any private burial place.

According to Shippen, only corpses protected under the auspices of religion and private property were exempt from dissection. Other (criminal, indigent, and anonymous) bodies warranted postmortem experimentation. Across the United States, black bodies were rarely permitted to remain long in their graves, if they were even buried at all. Tales of "night doctors" stealing cadavers and hastening the deaths of those in their care are embedded in African American folklore.[50] Slaveholders sold the corpses of deceased slaves to medical schools, continuing to profit from their bodies after death. In New York, African American burial grounds were routinely pillaged in the quest for specimens.[51] In 1788, the city's free black community petitioned the New York Common Council to prevent the invasion of their cemeteries by "students of physick": "Under the cover of night, and in the most wanton sallies of excess, [they] dig up the bodies of the deceased friends and relatives of your petitioners, carry them away and without respect to age or sex, mangle their flesh out of a wanton curiosity and then expose it to the Beasts and Birds."[52] Despite the trauma embedded in this rhetoric, the petitioners tempered their request. Presuming that a ban on resurrecting black bodies was unattainable, they acknowledged "the necessity of physicians and surgeons consulting dead subjects for the benefit of mankind." Echoing Shippen's defense of anatomy as "useful to Mankind," the petitioners asked only that dissections be "conducted with that decency and propriety which the solemnity of such an occasion requires."[53] Regardless of this measured logic, the council ignored their plea. As long as dissection was reserved for outcast cadavers, those of "legitimate" citizens were perceived safer in their graves. As one New Yorker wrote that same year, "the only subjects procured for dissection are the productions of Africa, and if those characters are the only subjects of dissection, surely no person can object."[54] It was not until grave robbers exhumed "respectable" bodies that public outrage reached a fever pitch. Just prior to the 1788 riot, the *New York Packet* reported: "The interments not only of strangers and the blacks had been disturbed, but the corpses of some respectable persons were removed."[55]

In nineteenth-century Philadelphia, the incessant snatching of bodies from the almshouse graveyard scandalized local residents. The Board of Guardians failed to protect the dead in their charge from the thriving enterprise of resurrection. In 1845 several members petitioned the board to prevent body snatching, arguing that fear of dissection endangered the health of residents: "That it occasions dread and anxiety in the minds of some of the inmates of this House, is a well known fact." Inhabitants knew that burial within the almshouse cemetery was a farce when the medical colleges were in session, "and to be buried elsewhere is some times asked as the last and greatest favor."[56] The board rejected the petition, insisting that "the colleges must have subjects," and if doctors were turned away, they would plunder church cemeteries and private burial grounds.[57] A decade later, in 1856, an investigation into the sale of bodies from the Philadelphia Almshouse found that Dr. E. B. Mosley, a member of the board, "prostituted his office to his own personal profit in making merchandise of the bodies of deceased paupers." In the scandal's aftermath, the Board of Guardians became known as the "Board of Buzzards."[58]

In 1854, one year prior to the first edition of *Leaves of Grass* and sixty-five years after the initial Anatomy Act of 1789, New York State passed the Act to Promote Medical Science and Protect Burial Grounds, commonly known as "the bone bill," which allowed for the donation of "unclaimed" corpses to medical schools.[59] This anatomical appropriation was less benign than it sounds. As Delaree's story exemplifies, these were not unmourned bodies. The status "unclaimed," when applied to a cadaver, was primarily an economic and social designation. Dissection was no longer a fate fit only for executed criminals; it could now be legally inflicted on any corpse whose family could not afford to pay for burial (although, in practice, this was happening all along). By the late nineteenth century, many states had passed laws requiring public institutions such as almshouses, prisons, and asylums to deliver cadavers to medical schools if those bodies would otherwise be buried at public expense. In Whitman's America, only those with financial means could hope to remain untouched by the anatomist's scalpel.[60]

The "bone bill" failed to alter the sources of cadaver supply. On the contrary, the legislation was designed to ensure that anatomical material would continue to be sourced from America's poorest citizens.[61] As Michael Sappol has established, institutionalization already signified a form of "social death."[62] Anatomy acts enhanced this humiliation by subjecting marginal bodies to the indignity of dissection, previously seen as the aftermath of capital punishment. Dissected corpses were denied the hospitality of burial and publicly wounded in the anatomist's theater, a fate that was thought to have dire consequences in the afterlife. Belief in the marriage of flesh and spirit for a liminal period after death and in the biblical resurrection fueled resistance to dissection. The stark disparity between posthumous violence and the "beautiful deaths" represented in sentimental literature rendered dissection a horror reserved for bodies that were considered expendable, in this life and the next.[63]

Sappol has argued that the boundaries between anatomist and cadaver not only reinforced the Cartesian divide between mind and matter, but also reflected antebellum social stratification:

> In this cultural poetics, the *dissector*, the generator of meaning, was identified with *mind*; the *dissected*, those whose bodies were appropriated as the medium through which meanings were generated, were identified with *body*. . . . The anatomist, recruited from the middle and upper echelons of society, served as an iconic representation of *spirit*. The cadaver, conscripted from ranks of black people, criminals, prostitutes, the Irish, "freaks," manual laborers, indigents, and Indians, served as an iconic representation of *matter*.[64]

Whitman insists upon the divinity of matter, celebrating his own flesh and that of countless others with whom he "merges."[65] His democratic catalogs render compassionate portraits of the "outcasts" whose cadavers were the primary subjects of anatomical investigation, including slaves, Native Americans, prostitutes, the insane, opium addicts, suicides, the disabled, and the impoverished.[66] For Whitman, the human body and the immortal soul are inextricably welded:

Clear and sweet is my soul . . . and clear and sweet is all that is not my
soul.
Lack one lacks both . . . and the unseen is proved by the seen,
Till that becomes unseen and receives proof in its turn.
. . .
Welcome is every organ and attribute of me, and of any man hearty and clean,
Not an inch or a particle of an inch is vile, and none shall be less familiar
than the rest.[67]

The soul is "clear and sweet," as is "every organ" and "particle" of the body. The absence of either entity equates to the absence of both. The existence of the "unseen" spirit is "proved" by the visible human body. When the physical body "becomes unseen" in death, the soul attached to it attains "proof" of divinity. To return to Sappol's analysis of the surgeon as spirit and the cadaver as matter, what would such a designation have wrought for Whitman, given his faith in spiritual and material unification? How can the mind dissect the body?

Whitman's fidelity to the body's divinity led him to adopt ambivalent attitudes toward the emerging science of death.[68] While vehemently opposing body snatching, the poet recognized the medical advancements that anatomy promised. He was able to divorce resurrectionism from the science underpinning the market for stolen bodies. In the preface to the 1855 edition of *Leaves of Grass,* he included anatomists alongside astronomers, spiritualists, chemists, mathematicians, phrenologists, and historians as members of "the sinewy race of bards" who "underlie the structure of every perfect poem."[69] In the poem eventually titled "I Sing the Body Electric," Whitman unites the "skin trades" of anatomy and slavery to catalog the evolutionary "wonders" of the human form:[70]

Gentlemen look on this curious creature,
Whatever the bids of the bidders they cannot be high enough for him,
For him the globe lay preparing quintillions of years without one
animal or plant,
For him the revolving cycles truly and steadily rolled.
. . .
Examine these limbs, red black or white . . . . they are very cunning in
tendon and nerve;
They shall be stript that you may see them.

Exquisite senses, lifelit eyes, pluck, volition,
Flakes of breastmuscle, pliant backbone and neck, flesh not flabby,
    goodsized arms and legs,
And wonders within there yet.

Within there runs his blood . . . . the same old blood . . . the same red
    running blood;
There swells and jets his heart . . . . There all passions and desires . . .
    all reachings and aspirations:
Do you think they are not there because they are not expressed in
    parlors and lecture-rooms?[71]

The specimen body of the "slave at auction" demonstrates the universality of human divinity: "If life and the soul are sacred, the human body is sacred."[72] The slave's body is dissected and articulated to reveal "wonders within." The limbs are "stript" so that the reader can admire their "cunning tendons." Paradoxically, Whitman's poetic vivisection seeks to restore human value by cataloging the inherent *sameness* of internal organs and "exquisite senses." Blood runs red beneath the skin of "red, black, or white" bodies. Yet this anatomical demonstration exists within a flawed economy of racial hierarchy. Whatever price is paid for this "curious creature" will never be "high enough" to compensate for the "revolving cycles" of evolutionary time that created him. At the poem's close, the structural hypocrisy of antebellum America's body-snatching phobia is rendered in black and white: juxtaposing the atrocities of slavery and the posthumous violence of dissection, Whitman insists that those who "degrade" "living human[s]" are as damned as those who "defile" the dead: "Who degrades or defiles the living human body is cursed, / Who degrades or defiles the body of the dead is not more cursed."[73]

Despite Whitman's sympathetic treatment of African Americans in his poetry, he did not advocate equal rights or citizenship for black Americans following the abolition of slavery. He was first and foremost a Unionist, and he subscribed to nineteenth-century theories of racial superiority that are nothing short of appalling to contemporary readers.[74] In the *Brooklyn Times* (1858) Whitman wrote: "Who believes the Whites and Blacks can ever amalgamate in America? Or who wishes it to happen? Nature has set an impassable seal against it."[75] While marginal cadavers were incorporated as raw material by nineteenth-century anatomists, Whitman incorporated "outcast" bodies toward very different ends. The specimen is not a dehumanizing tool for Whitman, but a model of collective identity. Anatomical symmetry reveals our shared humanity. Regardless of race, all blood runs red, all organs are equally "lifelit." Yet, there are ways in which his mode of resurrecting the socially dead betrays its own rapacity. This speaks to the dual valence of anonymity: it can be considered an ethical good because Whitman claims to accept all, but often his "outcasts" are so anonymous as to lose all specificity. Does his capaciousness risk replicating the avarice of grave robbers? Does anonymity reduce marginal bodies to use-values? Is the "poet of the body" also guilty of anatomical appropriation?[76] Borrowing from both anatomists and their detractors, Whitman aspires to become a "resurrectionist" in another, more democratic sense, absorbing and reviving the dead through the "common" grass that "bathes the globe."[77] While he does not entirely achieve these aims, Whitman absorbs shifting incarnations of America and her citizens, bound together under an optimistic "flag," "out of hopeful green stuff woven."[78] His poetic democracy incorporates the diversity of bodies that inhabit it; living and dead, they are irrevocably tied to the earth.

The Uncut Hair of Graves

Whitman inhabited a cityscape grappling with environmental waste and the public abandonment of human bodies. As we have seen, burial rituals were largely denied to the urban poor. Many died publicly in almshouses, asylums, or indigent hospitals. Even the corpses of those who died at home received scarce attention unless they were fortunate enough to have relatives who could afford funeral costs. In most cases, the body was unceremoniously transported to a potter's field. Burial workers found it difficult to keep up with rising death tolls. During the cholera epidemics, corpses remained uncollected in the streets for days before being buried in trench graves.[79] The spectacle of urban decay and its sensory consequences was impossible to ignore. As Maria Farland has observed, "*Leaves of Grass* confronts the question of whether this escalating demographic density—and the resulting waste and decomposition, could be absorbed by the natural environment."[80]

Exhumations of the dead were not only medically, but also environmentally, motivated. The 1845 Sanitary Commission's report cautioned against the pathological "influence of grave yards, vaults, and other burial places in large cities, upon the health of the inhabitants."[81] Religious treatises such as John Brazer's "On the Burial of the Dead" (1861) encouraged the relocation of cemeteries outside city limits.[82] Whitman's own mid-century prose responds to acute anxieties surrounding the decomposing matter that littered New York City streets.

The poet described a cityscape drowning in "receptacles of filth. . . . Introducing disease and death into the systems of those who use it."[83] The chaotic experience of urbanity that was central to *Leaves of Grass* was simultaneously destroying the landscape Whitman also celebrated. The same year that the first edition was published (1855) marked a pivotal moment in the history of medicine: a formal commission was appointed to investigate sanitary conditions within New York City, under the governance of the New York Academy of Medicine.[84] Prior to the Civil War, Whitman reconciled this urban environmental crisis through the metaphor of regeneration. Similarly, transplanting corpses to pastoral cemeteries alleviated widespread fears of posthumous contagions, allowing nature to cleanse the human detritus of urbanity.

Leaves of Grass was shaped by rapid changes in horticultural and burial practices. Whitman articulates environmental anxieties arising from a cityscape "impregnated with the impure drainings of graveyards."[85] The dangers of decay were believed to be airborne, capable of permeating stone masonry and infiltrating living bodies. Emerging biochemical theories offered an antidote to the miasmatic model of contagion. Influenced by the idea that pollution arising from human remains could be detoxified by flora, botanical regeneration underlies Whitman's antebellum poetics. In the poem eventually titled "Song of Myself," the libidinal surge that accompanies mourning is both internalized and externalized: buried within the earth, and absorbed into the bodies of those who inhabit it.

In "Song of Myself," Whitman seeks to capture a series of unutterable experiences, to put into words something outside the reach of language: the object "without name."[86] Yet this translation is inevitably partial. Certain elements remain "unspeakable": "[It] is a word unsaid, it is not in any dictionary or utterance or symbol."[87] Linguistic elusiveness lies at the heart of *Leaves of Grass*, alongside a meditation on the symbiotic reciprocity of burial. Through poetic "composting," corpses are transformed from "accumulations of filth" into "divine materials," recapturing the posthumous body as an object of desire.[88] The ascent of grass through the mouths of the dead parallels Whitman's anatomical and environmental aphasia. Through the cannibalistic magic of incorporation, Whitman's grass "tongues" swallow unspeakable losses.[89]

Whitman's hospitality toward the unburied dead archetypically recalls Antigone, who, as punishment for her subversive act of burying (and then reburying) her brother, was herself buried alive. Judith Butler has argued that Antigone knowingly embraces the death drive, understanding the consequences of her refusal to comply with Creon's edict that her brother's corpse remain untouched.[90] Butler's reading of Antigone speaks of an ethical responsibility toward lives that are deemed "ungrievable"—those that exist outside the parameters of "productive," societal mourning. Mourning the marginal other catalyzes a recognition of the self's liminality:

> The drama of reciprocal recognition begins when one consciousness finds that it is lost, lost in the Other, that it has come outside itself, that it finds itself as the Other or, indeed, *in* the Other. . . . Consciousness seeks a retrieval of itself, only to recognize that there is no return from alterity to a former self but only a transfiguration premised on the impossibility of return.[91]

Antigone mourns a beloved sibling. The love she bears her brother compels her to defy the tyrant's command that his body remain exposed to the elements. Yet it is equally transgressive to mourn the anonymous "corpses of outcasts" in the face of a hostile or indifferent state (as Whitman does throughout *Leaves of Grass*).[92] Selfhood is temporarily eclipsed in an attempt to comprehend the suffering of a dead stranger, who remains perpetually unknown. While no one could accuse Whitman of self-effacement, his persona "merges" with other bodies to catalog and translate their experiences.[93] Butler theorizes that transfiguration occurs when consciousness becomes indistinguishable from the absent Other. The psyche is altered not only by the recognition of this inherent difference, but also by the impossibility of returning to its prior status, untouched by otherness. Reciprocal mourning requires a borderless hospitality, unqualified by the limits of selfhood. One must be willing to transgress those boundaries with the knowledge that this trespass is permanent, that one will emerge forever changed.

Whitman constructs a discourse on the hospitality of mourning strangers, those collectively denigrated by economic, racial, or sexual otherness. Throughout the fluctuating editions of *Leaves of Grass*, the poet demonstrates a reverence for marginal bodies, those who were denied burial rites and became unwilling subjects of anatomical investigation. "Song of Myself" unites a series of eco-erotic requiems that seek to translate and transplant melancholic attachments to the anonymous dead. Whitman creates a libidinally charged landscape where the "grass of graves" acts as a conduit for pastoral transcendence:[94]

> Tenderly will I use you curling grass,
> It may be you transpire from the breasts of young men,

It may be if I had known them I would have loved them;
It may be you are from old people and from women, and from offspring taken soon out of their mothers' laps,
And here you are the mothers' laps.[95]

Due to its human origins, the grass must be used "tenderly." Floral anthropomorphism allows the poet to vicariously touch the "uncut hair" of unknown "dead young men." Grass penetrates their disintegrating bodies, passes through their mouths, and surfaces beneath the poet as a signifier of morphological change:

This grass is very dark to be from the white heads of old mothers,
Darker than the colorless beards of old men,
Dark to come from under the faint red roofs of mouths.[96]

Darkness contrasts with the "faint" red mouths and "white" hair of the dead to create an uncanny fertility. Whitman's sexuality is embedded within this eco-erotic structure, inextricably linked, through processes of decay and regeneration, with mourning the unknown. Human particles decompose and reemerge entirely altered, dispersed through the earth and finally incorporated into living bodies.

The ascent of grass through dead mouths inverts psychoanalytic theories of mourning. Freud initially conceptualized melancholic incorporation as resistance to the finality of death—refusing to conclude the processes of mourning.[97] Rather than integrating the lost other, the subject absorbs the loss itself, creating an intrinsic homage to trauma that lives within the survivor. Working primarily with Holocaust survivors in the aftermath of World War II, the psychoanalysts Nicholas Abraham and Maria Torok expanded Freud's theory of melancholia as unfinished (or unfinishable) grief.[98] Abraham and Torok divided "internalization," the primary mechanism behind mourning, into two subsets: "introjection," the process of symbolically absorbing the other in "successful" mourning; and "incorporation," the fantastic wound of melancholia in which the loss remains unrealizable and the other is encrypted within the psyche.[99] Their "exquisite corpse" represents a lost erotic attachment that is "buried like a corpse" within the mind, only to be resurrected by the "libidinal surge" that accompanies the actual death of a beloved other. Partial mourning generates linguistic barriers known as crypts, in which the ego incorporates the object-loss, rather than the object itself. The crypt-bearer is rendered mute in relation to this loss. To utter any word that alludes to it would be a catastrophic acknowledgment that the mourned other is not alive within, but dead without.

> Because our mouth is unable to say certain words . . . we fantasize . . . that we are actually taking into our mouths the unnameable, the object itself. . . . Failing to feed itself on words . . . the mouth absorbs in fantasy all or part of a person—the genuine depository of what is now nameless.[100]

When it cannot "feed" on words, the melancholic mouth consumes death instead, transforming an external void into an internal wound. The "exquisite corpse" is psychically interred through the "fantasy of incorporation," "introducing all or part of a love object or a thing into one's own body."[101] Whitman inverts this psychological process: rather than a live mouth swallowing a dead object, a live object passes through a dead mouth. The key to why (and how) this inversion takes place can be found through an exploration of nineteenth-century biochemical and botanical theories.

Central to an understanding of the sanitation panic surrounding urban burial is the antebellum perception that decay resulted in the excretion of miasma: a "noxious vapour rising from putrescent organic matter . . . which pollutes the atmosphere."[102] Prior to germ theory, miasma was thought to be the primary carrier of infectious diseases. Whitman believed that decomposing matter "generates a kind of *miasma* in the air," after which "disease and fevers follow."[103] Medical literature warned against the dangers that overcrowded graveyards posed to public health. Francis D. Allen's *Documents and Facts, Showing the Fatal Effects of Interments in Populous Cities* (1822) cautioned residents against living in proximity to burial grounds, arguing that the "putrid exhalations arising from grave-yards" penetrated stonework to invade human bodies.[104] In 1823, a member of the Massachusetts Medical Society argued that decomposition would cause "the earth of cemeteries" to become "so filled with putrid matter and effluvia, as to endanger the health and the life of all those exposed."[105] The Rural Cemeteries Act (1847) promoted the construction of pastoral graveyards throughout New York, influenced by the popular belief that air tainted by miasma could be purified by circulation through leaves and trees.[106]

Garden cemeteries such as Brooklyn's Greenwood are a horticultural legacy of the urban sanitation movement.[107] In her literary analysis of rural cemeteries, Desirée Henderson establishes their dependence on metaphysical and political philosophies: "cemetery literature promises the universal potential of a transcendent awakening, a democratic ideal that permeates every element of the cemeteries in both design and depiction."[108] These landscaped graveyards on the peripheries of cities emerged not only from pastoral impulses, but also from more practical concerns regarding environmental contamination and body snatching. In 1831 Jacob Bigelow, a physician from Boston, initiated the construction of Mount Auburn, the first rural cemetery in the United States. Merging interests in botany with a desire to initiate sanitary burial practices, Bigelow designed the cemetery as a pastoral idyll for the living and the dead. The philosophical impetus behind the establishment of ecosystemic barriers between cemeteries and cities was directly correlated with emerging scientific theories on the relationship between decomposition and chemistry.

As David S. Reynolds has established, Whitman was influenced by the German chemist Justus Liebig, the founder of agricultural chemistry. When the English translation of Liebig's *Chemistry in Its Application to Physiology and Agriculture* (1842) was published, Whitman wrote a glowing review in the *Eagle*: "Chemistry involves the essences of creation, and the changes, and the growths, and the formations and decays of so large a constituent part of the earth."[109] Liebig gave credibility to the theory of biological continuity through chemical transference. He hypothesized that as an organism decomposed, its atoms chemically reunited, creating "another arrangement of the atoms of a body, that is, to the production of a compound which did not before exist in it."[110] Any diseases the organism may have carried were lost in this transference: "the miasms and certain contagious matters [that] produce diseases in the human organism" become "not contagious when the organism is absorbed into the earth."[111] The democratic unity inherent in this anatomical exchange was not lost on Whitman. Liebig's belief that "the active state of the atoms of one body has an influence upon the atoms of a body in contact with it" was resurrected in the opening stanza of *Leaves of Grass*: "every atom belonging to me as good belongs to you."[112] Cellular unification negates the finality of death:

> They are alive and well somewhere;
> The smallest sprout shows there really is no death,
> And if there ever was it led forward to life, and does not wait at the end to
>   arrest it,
> And ceased the moment life appeared.
>
> All goes onward and outward. . . . and nothing collapses,
> And to die is different from what any one supposed, and luckier.[113]

Within Whitman's landscape of corporeal, spiritual, and ecological unification, each organism embodies divinity and all things are interconnected: "They are but parts . . . any thing is but a part."[114] *Leaves of Grass* adheres to the Emersonian tenet of interrelation, "each particle is a microcosm," but for Whitman these particles continually unite until they become almost interchangeable.[115] Ecological synecdoche reveals the intrinsic continuity of all organisms. Through regenerative morphology, grass merges with the "hair" of the dead, who are far "luckier" than the living can fathom.

So Many Uttering Tongues

"The lack of any words," Whitman writes in *An American Primer*, "is as historical as the existence of words. As for me, I feel a hundred realities, clearly determined in me, that words are not yet formed to represent. . . . What is not said is just as important as what is said, and holds just as much meaning."[116] The impossibility of representation is reflected in the authorial presence, who resists linguistic confines to retain his singularity: "I, too, am untranslatable."[117] Embodying the interconnectivity of flesh and spirit, Whitman gives voice to unspeakable traumas:

> I am the poet of the Body and I am the poet of the Soul,
> The pleasures of heaven are with me and the pains of hell are with me,
> The first I graft and increase upon myself, the latter I translate into a
> new tongue.[118]

While the ecstasies of heaven are seamlessly "graft[ed]" onto the poet's body, pain defies language and must be "translat[ed] into a new tongue." Throughout *Leaves of Grass*, unutterability is represented by cellular decomposition. The hospitality of burial enables both absorption into, and dispersion throughout, the ecoerotic landscape. As a vehicle for the process of merging with the dead and with the earth itself, this decomposition involves not the cessation of being, but the height of it: decay as transcendence. Incorporation erases the need for language; once the body is (dis)integrated there is no need for the particularity of a name.

Whitman's grass mouths reflect biochemical processes that are perceived as both infinite and "unnameable." In "A Plea for Water" (1851), Whitman urged readers to "imagine all the accumulations of filth in a great city . . . the unnameable and immeasurable dirt that is ever, ever filtered into the earth, through its myriad pores."[119] The following image of botanical rapaciousness from James F. Johnston's *The Chemistry of Common Life* (1855) illuminates the incorporative function of Whitman's porous leaves:

> Over the surface of these leaves are sprinkled countless pores or mouths, which are continually employed in separating and drinking carbonic acid gas. . . . A common lilac with a million of leaves, has about four hundred thousand million of pores or mouths at work, sucking in carbonic acid; on a single oak tree, as many as seven millions of leaves have been counted.[120]

Whitman's botanical orality is a vehicle for swallowing loss and reviving the dead. "Limitless leaves" containing their own multiplicity of mouths pass through the mouths of corpses.[121] Biochemical morphology endows grass with erotic and psychological significance. In the process of imbibing decay, Whitman's grass tongues resurrect the lost object: "The smallest sprout shows there really is no death." The act of swallowing loss is far from silent. Whitman's grass mouths speak to the ecoerotic possibilities of regeneration: "O I perceive after all so many uttering tongues! / And I perceive they do not come from the roofs of mouths for nothing."[122] This is the language of an unconscious haunting, the tongue that translates itself. In denying the finality of death and the silence of the dead, Whitman refuses to allow the absence of corporeality to negate the erotic potential of an entity. These poetic tongues speak of a democratic afterlife, in which each soul must be "considered," and none can "fail." Fidelity toward the dead, known and unknown, is assured:

> Each who passes is considered, and each who stops is
>   considered, and not a single one can it fail.
>
> It cannot fail the young man who died and was buried,
> Nor the young woman who died and was put by his side,
> . . .
> Nor any thing in the earth, or down in the oldest graves of the earth,
> Nor any thing in the myriads of spheres, nor one of the myriads
>   of myriads that inhabit them,
> Nor the present, nor the least wisp that is known.
> . . .
> What is known I strip away. . . . I launch all men and women
> forward with me into the unknown.[123]

Whitman elegizes the victims of epidemic disease, insanity, and poverty whose bodies have been "slaughtered and wrecked." He questions society's ethical and environmental responsibilities toward the human body, regardless of class, gender, race, or productivity. The poet's faith in ecological hospitality is so complete that he seeks to "strip away what is known." This erasure of identifying markers renders all men and women as strangers. In the "oldest graves of the earth," social hierarchies vanish with the bodies' collective decay. This desire to embrace unknowability contrasts starkly with the anxiety surrounding the anonymity (and infinity) of the dead that pervades Whitman's work during and following the Civil War. Postbellum editions of *Leaves of Grass* often portray the earth's incorporative capacity as exhausted by the "numberless" slaughtered whose blood "saturated" the landscape.[124]

In the antebellum editions, Whitman's primary mechanism for "contain[ing] multitudes" is to "merge" with others, vicariously absorbing their diverse experiences. The poet moves as a "free companion" through the landscape, "swallow[ing]" its inhabitants:[125]

> All this I swallow and it tastes good. . . . I like it well, and it becomes mine,
> I am the man. . . . I suffered. . . . I was there.
> The disdain and calmness of martyrs,
> The mother condemned for a witch and burnt with dry wood, and her
> children gazing on;
> The hounded slave that flags in the race and leans by the fence, blowing
> and covered with sweat,
> The twinges that sting like needles his legs and neck,
> The murderous buckshot and the bullets,
> All these I feel or am.[126]

Paradoxically, suffering "tastes good"; the poet experiences heightened arousal following its consumption. His pleasure is both sadistic (in the sense that it originates within the body of the Other) and masochistic (in the sense that pleasure does not arise from inflicting pain, but rather from the sensory experience of the other's pain as one's own):

> Agonies are one of my changes of garments;
> I do not ask the wounded person how he feels . . . . I myself become the
> wounded person,
> My hurt turns livid upon me as I lean on a cane and observe.[127]

Whitman disguises his presence in the "agonies" of others as transient "garments." Although he "become[s] the wounded person," this condition is impermanent: otherness can be cast aside. Yet, his assumed "agonies" turn "livid" upon their observer, demonstrating an awareness of the limits of appropriation. Whitman's permeability is fundamentally empathic, allowing him to (temporarily) inhabit the bodies of others, to swallow and then to translate their experiences. In this sense, the poet is the human equivalent of Liebig's rapacious leaves, absorbing trauma rather than miasma.

The Whitmanian phenomenon of "merging" parallels spiritualist practices of disembodied communication. Beginning in 1848, spiritualism soared in popularity as practitioners sought ongoing dialogues with the dead. Bodily transcendence was a central tenet of the faith, which offered adherents the chance to escape the constraints of their material forms.[128] In a July 1862 letter to *The Banner of Light,* a popular spiritualist newspaper, a reader inquired as to whether it was "possible for the spirit of the medium to commune with friends when apart from her own body?"[129] The editor's response demonstrates an uncanny resemblance to the Whitmanian "merge": "It is possible. Notwithstanding there is a sympathy kept up between spirit and body, yet the spirit itself is free to go wheresoever it wills; free, it finds conditions adapted to its use, to employ them any time or place, however distant."[130]

Whitman observed the spiritualist movement with curiosity, and, at least initially, placed it within the same scientific category as anatomy and chemistry.[131] In a self-authored review of the 1855 edition, the poet described himself as "the true spiritualist. He recognizes no annihilation, or death, or loss of identity."[132] In Whitman's mind, the aims of the poet and those of the medium were inherently similar: to act as magnetic forces for the attraction of spirits that defy death. As Molly McGarry has established, Whitman's own "forays into Spiritualism provide crucial context for understanding [his] urge to forge intimate bonds with spiritual and embodied strangers."[133]

Mediumship is, at its heart, an invitation. The medium must be willing to be entirely occupied by the ghost. Derrida invites us to consider a hospitality that extends beyond death, an invited haunting: "We need to go further and think of hospitality toward death. There is no hospitality without memory. A memory that does not recall the dead person and mortality would be no memory. What kind of hospitality would not be ready to offer itself to the dead one, to the revenant?"[134] Haunting and hospitality are irrevocably linked as mnemonic devices that recall and resurrect the lost other. Both require a host in order to stage their arrival. As Anne Dufourmantelle writes, "The hostis responds to hospitality in the way that the ghost recalls himself to the living, not letting them forget."[135] Swallowing the suffering of others, Whitman captures the tension between the turning away of survival and the embrace of death. Yet Whitmanian incorporation extends beyond inviting the other to enter his body. The merging poet is himself disembodied—a permeable entity who passes through the flesh of others. To take on the "impossible mourning" that the revenant dictates is to become a specter oneself.

According to Abraham and Torok, the need for incorporation necessitates the cannibalistic magic of swallowing the dead: "The fantasy of incorporation simulates profound psychic transformation through magic, it does so by implementing literally something that has only figurative meaning. So in order to 'swallow' a loss, we fantasize swallowing (or having swallowed) that which has been lost, as if it were some kind of thing."[136] Whitman's lost objects are the anonymous strangers he melancholically desires: "if I had known them I would have loved them." In the absence of a name, it is necessary to fill the mouths of the dead with grass, which penetrates them as the poet no longer can.

Judith Butler's work on Antigone identifies correlations between unutterability, desire, and burial. Resisting the Freudian pathology of disordered mourning, she describes melancholia as the psychological defense of a love that transcends the taboos erected to thwart it:

> What is produced is a shadowy realm of love, a love that persists in spite of its foreclosure in an ontologically suspended mode. What emerges is a melancholia that attends living and loving outside the liveable and outside the field of love, where the lack of institutional sanction forces language into perpetual catachresis.[137]

Melancholics infinitely mourn losses that exist outside "institutional sanction" and linguistic representation. The queerness that operates between unnameable love and ontological instability recurs throughout *Leaves of Grass*. Michael Moon reflects on Whitman's resistance to "thinking of the deaths of others as the making deficient of our own bodies," and of "death as absolutely rupturing the possible erotic relation of a living person to a dead one."[138] Whitman's desire for the dead manifests through ecoerotic projection, transcending the physicality of the sex act to locate arousal somewhere outside of corporeality. Grass symbolizes the passage from abjection to ecstasy, the infinite possibilities of regeneration.

Torok describes the "libidinal invasion" that often accompanies the onset of mourning as "a widespread, if not universal phenomenon."[139] A sexualized, spiritual invasion immediately prefaces Whitman's meditation on the "grass of graves."[140] In the preceding lines, the poet acts as medium, invoking the "self" possession of his own soul.[141] He entreats the soul to "loafe with me on the grass" and invites the "hum" of its "valved voice" to speak through him. The poet recalls the "transparent summer morning" when his soul accepted this invitation:

> You settled your head athwart my hips and gently turned over
>   upon me,
> And parted the shirt from my bosom-bone, and plunged your
>   tongue to my barestript heart,
> And reached till you felt my beard, and reached till you held
>   my feet.[142]

Whitman's heart is "stript" by the soul's tongue, just as the mouths of the dead are pierced by tongues of grass. Torok describes the erotic "surge" that occurs at the moment of loss as a final attempt at unification with the departing other: "The ego makes use of the [lost] object . . . to achieve its libidinal awakening."[143] The catalyst for Whitman's transcendence is this ecoerotic encounter with his own soul, which resurrects the bodies of the unknown dead as blades of grass. Decomposition and regeneration enable the lost other's revival:

> Smile O voluptuous coolbreathed earth!
> Earth of the slumbering and liquid trees!
> Earth of the departed sunset! Earth of the mountains misty-topt!
> Earth of the vitreous pour of the full moon just tinged with blue!
> Earth of shine and dark mottling the tide of the river!
> Earth of the limpid gray of clouds brighter and clearer for my sake!
> Far swooping elbowed earth! Rich apple-blossomed earth!
> Smile, for your lover comes![144]

More than an inscription or personification of homoerotic desires upon the landscape, Whitman's ecoeroticism derives sexual gratification from nature.[145] Fertility emerges from decay and the earth itself becomes an object of arousal. Following his union with the soul, Whitman is "thrust" "tight" to the earth, upon which the poet inscribes his desire for bodies living and dead, including his own.[146]

In the end, Whitman's ecoerotic elegy comes full circle. To complete his final merge, the authorial body must disintegrate, a rupture foreshadowed by his recurrent possession of other bodies. The poet casts aside his own flesh, a final abandonment of corporeal garments: "I depart as air, I shake my white locks at the runaway sun, / I effuse my flesh in eddies and drift it in lacy jags."[147] Whitman's effusion of self performs a desire to reunite with the dead at the expense of his own body, in keeping with Abraham and Torok's observation that melancholics "lend their own flesh to the phantom object of love."[148] Whitman's airborne specter wields a miasmatic permeability. Etymologically, effusion suggests the "pouring out or spilling of liquid," sending forth "air, heat, light, or odours," and finally "the pouring out of blood by a wound."[149] Whitman transforms into a substance capable of cellular dissemination via blood or air—both physical and phantasmal:

> I bequeath myself to the dirt to grow from the grass I love,
> If you want me again look for me under your bootsoles.
>
> You will hardly know who I am or what I mean,
> But I shall be good health to you nevertheless,
> And filter and fibre your blood.
>
> Failing to fetch me at first keep encouraged,
> Missing me one place search another,
> I stop somewhere waiting for you.[150]

As with Antigone, the consequence of Whitman's desire for the dead is the live-burial of the self. This authorial ghost inverts the logic of hospitality: the poem closes with an awaited visitation, in which the guest patiently anticipates the arrival of the host. Whitman's is a symbiotic haunting: the ghost nourishes his host, promising "good health." The authorial corpse remains biologically enmeshed with the living, lingering in the bodies of his descendants to "filter and fibre" their blood. Whitman's textual incorporation at the close of this first version of "Song of Myself" foreshadows the dispersal of poems throughout years of unrelenting revisions, as the poet widened his textual "memorial" to include countless "unfound soldiers."

Strange Collections

In the aftermath of the Civil War, the fourth (1867) edition of *Leaves of Grass* introduces a poem set upon the gates of a morbid threshold reminiscent of "The Tomb Blossoms." The scene has shifted from a rural cemetery to an urban hospital. "The City Dead-House" elegizes a prostitute abandoned outside the morgue:

By the city dead-house by the gate,
As idly sauntering wending my way from the clangor,
I curious pause, for lo, an outcast form, a poor dead prostitute brought,
Her corpse they deposit unclaim'd, it lies on the damp brick pavement,
The divine woman, her body, I see the body, I look on it alone,
That house once full of passion and beauty, all else I notice not,
Nor stillness so cold, nor running water from faucet, nor odors morbific
impress me,
But the house alone—that wondrous house—that delicate fair house—
that ruin!
That immortal house more than all the rows of dwellings ever built!
Or white-domed capitol with majestic figure surmounted, or all the old
high-spired cathedrals,
That little house alone more than them all—poor, desperate house!
Fair, fearful wreck—tenement of a soul—itself a soul,
Unclaim'd, avoided house—take one breath from my tremulous lips,
Take one tear dropt aside as I go for thought of you,
Dead house of love—house of madness and sin, crumbled, crush'd.
House of life, erewhile talking and laughing—but ah, poor house, dead
even then,
Months, years, an echoing garnish'd house—but dead, dead, dead.[151]

The position of this "unclaim'd" female body on the border between life (the city) and death (the morgue) mirrors her status on the outskirts of society. Like Delaree's husband, her cadaver is cast aside with "the careless indifference which is shown to the corpses of outcasts."[152] The dual designation of "unclaimed" and "prostitute" meant that this woman was almost certainly fated for dissection on "the ghastly tables of the dead house."[153] Whitman echoes the exact terminology used in the Act to Promote Medical Science and Protect Burial Grounds (1854) to legalize the dissection of unclaimed corpses. By the time Whitman published "The City Dead-House" in 1867, the "bone bill" had been legislation for thirteen years.

The sentimental domesticity (and privacy) necessary to achieve a "good death" contrasts starkly with the duality of "avoided houses" in the poem: the prostitute's cadaver abandoned on the dead-house steps. The glaring absence of mourning rituals is heightened by Whitman's rendering of the prostitute's body as a vacant "tenement of a soul—itself a soul." The word "tenement" reflects poverty and transient domesticity. Her body does not just contain the soul—it is "itself a soul." Society's pre- and postmortem neglect is the source of the woman's "ruin"; this is not a moral punishment for her status as a sex worker. Her "wondrous" body, once a manifestation of "passion and beauty," is now a posthumous spectacle.

"The City-Dead House" is most likely located either at Bellevue Hospital in Manhattan, or Kings County Hospital in Fort Greene.[154] One of the oldest hospitals in the nation, Bellevue began as an almshouse founded by the Dutch surgeon Jacob Hendrickssen Varrenvanger in 1658. Over the next century and a half, the hospital underwent massive expansions, architecturally and medically. In 1857 Bellevue was renovated to include a pathology center where surgeons performed autopsies and experiments rarely undertaken in private hospitals. By the 1870s, the hospital had become a vast network of charity wards, laboratories, and a morgue that met the pathological needs of the adjacent Bellevue Medical College.[155] Kings County Hospital also evolved from a local almshouse infirmary, gradually expanding to include an insane asylum and a medical research facility. Whitman was intimately familiar with Kings County Hospital, both personally and professionally. As editor of the *Brooklyn Daily Eagle*, he wrote a series of articles on the institution in 1846. In 1864 his brother Jesse was committed to the asylum ward, where he died six years later.[156]

Nineteenth-century public hospitals were ostensibly charitable institutions. Due to the chronic shortage of medical care, hospital trustees wielded the authority to decide who merited admission. They privileged patients who were deemed morally deserving, such as widows, orphans, soldiers, and sailors. The so-called unworthy poor (including prostitutes, alcoholics, criminals, and unwed mothers) were turned away. As the medical historian Howard Markel observes, "the river of human pathology at Bellevue had no end, and its sources were the slums and ghettos of New York. . . . From morning to late at night, year after year, the sick and needy pounded on the hospital's doors, begging for admission."[157] Given the location of both dead-houses within larger medical complexes, the prostitute's position at the morgue's gates symbolizes her exclusion from the hospital's care. Medical science was only interested in her corpse; the living woman would almost certainly have been refused entry.

James D. McCabe Jr.'s harrowing description of the Bellevue morgue in *Lights and Shadows of New York Life* (1872) illuminates the interior of "The City-Dead House":

> Over the lowest door of the front, on the upper side of Twenty-sixth street, is a single word in gilt letters—MORGUE. This door marks the entrance to the Dead House of New York, one of the most repulsive, but most terribly fascinating places in the city. . . . Bodies found in the streets or in the harbor are brought here for identification. They are kept a certain length of time, usually from twenty-four to forty-eight hours, and if not claimed by relatives or friends, are buried at the expense of the city. . . . The room is gloomy and cell-like in appearance. . . . The apartment is divided into two unequal portions by a partition of glass and iron. The smaller portion is used by the public. The remainder is devoted to the purposes of the establishment. Back of the glass screen are four stone tables on iron frames. . . . Stretched on these are lifeless naked forms, each covered with a sheet. A stream of cold water, from a moveable jet, falls over the lifeless face of each and trickles over the senseless forms, warding off decay until the last moment, in the hope that some one to whom the dead man or woman was dear in life will come and claim the body. A vain hope, generally, for but few bodies are claimed. Nearly all go to the potter's field. . . . On the walls back of the tables are suspended the clothing of the unfortunates, and of others who have preceded them. Maybe some friend will come along and recognize them, and the one who has been missing will be traced to this sad place. They form a strange collection, but they speak chiefly of poverty and suffering.[158]

The dead house is a magnetically abject space for both writers. McCabe finds it "one of the most repulsive, but most terribly fascinating places in the city." Whitman initially "pause[s]" at the gates because he is "curious," though his curiosity soon turns to melancholy as he observes the woman "deposit[ed] . . . on the damp brick pavement." McCabe's "stream of cold water" designed to "ward off decay until the last moment, in the hope that some one . . . will come to claim the body" mirrors Whitman's "stillness cold . . . running water from faucet." Like Whitman's "changes of garments," the discarded clothes of these "unfortunates" form a "strange collection" that "speaks" of "poverty and suffering." While the dead remain unknown, these spectral garments reflect the traumas of their late occupants to those who pause long enough to look.

In 1856 Whitman wrote a series of articles collectively titled under the anatomical pun "New York Dissected" in which he sketched a far less compassionate portrait of Manhattan's streetwalkers: "Dirty finery, excessively plentiful; paint, both red and white; draggle-tailed dress, ill-fitting; coarse features, unintelligent; bold glance, questioning, shameless, perceptibly anxious; hideous croak or dry, brazen ring in voice; affected, but awkward, mincing, waggling gait. Harlot."[159] This cruel depiction is a far cry from the poet's elegy for the "divine woman" discarded outside the morgue. Like the unsettling expressions of white supremacism that disturb the Whitman archive, it is difficult to reconcile this image with his empathic poetics.[160] While Betsy Erkkila stresses the democratic inclusivity of Whitman's catalogs, she also warns that Whitman's sweeping mergers "could operate paradoxically as a kind of formal tyranny, muting the fact of inequality, race conflict, racial difference within a rhetorical economy of many and one."[161] Ivy G. Wilson has interrogated the precarious boundaries of Whitman's democratic voice:

> Whitman's famous maxim about containing multitudes has often been understood as a metaphor for the democratizing impulses of the nation, but the presence of African Americans—suspended between the material and the apparitional, as it were—within the poetic space of his verse and other writings complicates any understanding of how the US cultural and literary imagination seeks to contain or otherwise demarcate its citizens.[162]

The prostitute's corpse lies on the border between anatomical materiality and divine spirituality. In "New York Dissected" Whitman wields the writer's pen as a surgical scalpel, to examine and excise the urban contagion of prostitution. In "The City Dead-House" he views the woman's body from the vantage of the poet "of Equality" who turns no one away, and turns away from no one.[163]

Whitman was not alone in his contradictory responses to the world's oldest profession. Prostitution evoked libidinal dissonance throughout the nineteenth century. In his study of medicine and mortality in nineteenth-century Paris, Jonathan Strauss examines female prostitutes as paradoxical symbols of revulsion and arousal. "The prostitute exposes another facet of unspoken fantasy hidden in the violence of disgust: miasma, putrefying, abject and horrible, she was nonetheless an object of desire."[164] Like the toxic effusions of corpses, prostitutes were seen as threats to urban hygiene, spreading not only venereal disease, but also miasmas "exhaled by foul and wine soaked bosoms."[165] In this context, the abandoned corpse embodies a duality of environmental and democratic failures—an impoverished life on the streets subjected her to the city's unrelenting pollution, while an unclaimed death denied her the ecological detoxification of burial and regeneration.

Embedded in the social chaos following Lincoln's assassination, the 1867 edition of *Leaves of Grass* revised many of the antebellum poems to reflect postwar society, alongside six new additions, including "The City Dead-House." In the context of postbellum debates on civil rights, the abandoned female body signifies the exclusion of racially and sexually marginal citizens from democratic discourse. Luke Mancuso describes Whitman's elegy for the prostitute as a means of recovering visibility for democracy's unseen strangers:

> At the heart of this text, the issue of disposable persons in a flawed democracy is argued with as much rhetorical force as it was by the Radical Republicans in the houses of Congress, on behalf of African Americans; only Whitman's marginal figure is a dead prostitute. . . . Socially outcast, the body of the prostitute requires the intervention of the poet's speaker in order that she may be represented visibly, in a democracy in which many are invisible. . . . This compelling poem enacts a recovery of the rightful place of human solidarity among strangers.[166]

Throughout *Leaves of Grass*, Whitman defends the divinity of peripheral citizens whose bodies were considered materially disposable. The mourning of strangers was a revolutionary act, unrecognized and unregulated by antebellum society. Anonymous mourning was subversive because its subjects were social outcasts: dispossessed and exiled in life, dissected and dehumanized in death.

It is possible that "The City Dead-House" was conceived in the aftermath of the Battle of Fredericksburg. Given Whitman's lifelong habit of recycling paper and reusing old notebooks, it is impossible to establish a definitive chronology. Several drafts of this poem appear in Whitman's 1862 notebook alongside entries from his visit to Falmouth, Virginia.[167] The fourth edition of *Leaves of Grass* was also the first to include Whitman's war poems, *Drum-Taps.*

Whitman's entrance into the war scene was triggered in December 1862 when he read the name "G. W. Whitmore" on a casualty list. Fearing this was a misprinted reference to his brother, George Washington Whitman, Walt departed immediately for the front.[168] His pocket was picked on the train platform, and he arrived in the capital penniless. Undaunted, Whitman wandered the Washington hospitals for two days and nights, unable to afford a coach further south, but determined to visit each ward, in case George had already been evacuated from Fredericksburg. He recalled the experience as "three days of the greatest suffering I ever experienced in my life."[169] Finally, he encountered William O'Connor, an abolitionist writer and friend from Boston, who arranged for his passage on a military train heading to Virginia.[170] Whitman's experience, while traumatic, was far from unique. The wartime press sensationalized casualty lists, while battlefields and hospitals overflowed with unidentified bodies.

In Falmouth, Whitman found George alive and relatively unscathed (a shell fragment had punctured his cheek, but he suffered no lasting damage). The poet remained at camp for over a week, crystallizing the devotion to wounded soldiers that would dominate his life for the duration of the war, and well beyond.[171] Drafts of "The City Dead-House" are bordered by descriptions of the temporary field hospital at Lacy House and the makeshift graves that lined the surrounding garden:[172]

> The graves with slight board, rudely inscribed with the names,
> In front of the hospital, the dead brought out, lying there so still
> The piece of board, hastily inscribed with the name, placed on the
>   breast, to be ready,
> The squad at the burial, firing a volley over the grave.[173]

Whitman's notebooks juxtapose images of a prostitute abandoned outside an urban morgue with observations of a rural manor reconfigured as a war hospital. Traditional domestic mourning rituals are once again aborted and inverted. In "The City-Dead House" the anonymous female corpse becomes a ruined house, absent any mourners. At Lacy House war casualties invade the domestic home that has become a battlefield triage unit. Perhaps from its inception (and at least from its first publication) Whitman's elegy for this "unclaim'd" woman was attached to his grief for unknown soldiers. Their bodies are linked by anonymity, abandonment, and disposability:

> O my sick soul
> How the dead lie
> Some lie on their back with faces up and arms extended
> The wounded—the surgeons and ambulances
> O the hideous damned hell of war
> O there is no hell more damned than this hell of war
> O beautiful you men!
> O the beautiful hair clotted! The faces![174]

Like the woman's body on the dead-house steps, the beautiful, upturned faces of fallen soldiers lie temporarily in a threshold space (the garden surrounding

Lacy House), awaiting swift disposal rather than ritualized burial. Following one of the bloodiest battles of the war, surviving Union soldiers set about the exhausting task of burying their dead. Union casualties at Fredericksburg exceeded 12,000; approximately 9,600 were wounded (many of whom would die in the aftermath, either of disease or infection), 1,284 killed in battle, and 1,769 captured or missing in action.[175] The psychological trauma of burial detail at Fredericksburg was compounded by the physical difficulty of the work: soldiers had to dig through frozen earth that was nearly impenetrable. Burial often consisted of shallow graves dug with bayonets or fragments of exploded shells, or mass trenches that contained an uncertain number of soldiers, the massacre so severe that their bodies were indiscernible.[176] The "hell of war" created posthumous conditions similar to the "careless indifference" previously reserved for "the corpses of outcasts."[177]

Just a few months before he journeyed to Fredericksburg, Whitman's first analysis of medical specimens appeared in print. In March 1862 Whitman wrote an article on Broadway Hospital for the *New York Leader.* Anticipating the traumatic aphasia that haunts his war prose, Whitman opens with an apologia for his inability to adequately capture the hospital's atmosphere: "Though I have been there twenty times, I feel unable to do justice, even to this kind of account of it, which only aims to skim over the surface."[178] Reading like a preface to the war memoranda he would soon begin writing, Whitman described the hospital's approximately 400 patients as "illustrating nearly all sorts of diseases, except infectious ones and insanity." Most were "surgical cases, contusions, fractures, wounds, &c." In addition to stagecoach drivers, passengers, and pedestrians injured in local accidents, there were "a good many soldiers . . . from the Volunteer Regiments passing through New York."[179]

Whitman had long cultivated an intimate camaraderie with stagecoach drivers. Given the chaos of New York City streets, their occupation was highly dangerous. The poet often visited injured drivers at Broadway Hospital.[180] A young driver appears in the following passage from "Song of Myself":

> The lunatic is carried at last to the asylum a confirmed case,
> He will never sleep any more as he did in the cot in his mother's
> bedroom;
> The jour printer with gray head and gaunt jaws works at his case,
> He turns his quid of tobacco, his eyes get blurred with the
> manuscript;
> The malformed limbs are tied to the anatomist's table,

What is removed drops horribly in a pail;
The quadroon girl is sold at the stand . . . . the drunkard nods by the
  barroom stove,
The machinist rolls up his sleeves . . . . the policeman travels his
  beat . . . .
  the gate-keeper marks who pass,
The young fellow drives the express-wagon . . . . I love him though
I do not know him.[181]

This sweeping catalog includes several recurrent figures in Whitman's poetry and his biography: the insane brother committed to an asylum, the amputee, the journeyman printer, the slave on the auction block, the anonymous but beloved young man. Their disparate "voices" are united by the poet's chameleon-like presence. The erotic undercurrent of this psychological permeability is enhanced by Whitman's desire for the driver, echoing the poet's elegy for the anonymous "dead young men" he "would have loved."[182] Kenneth M. Price has argued that Whitman's poetics often relied on racial and sexual passing, "sometimes in a specifically racial sense, but more often through the creation of a shape-changing, identity-shifting, gender-crossing protean self at the heart of *Leaves of Grass*."[183] In addition to surveying diverse aspects of American life, Whitman's poetic catalogs represent a collection of representative bodies—specimens. Robert Roper has criticized the tendency to interpret Whitman's fascination with young men as "collecting," which he describes as a "cynical, straightforward reading." His analysis focuses on Whitman's cruising lists as working through his theory of adhesiveness.[184] While I agree that Whitman's notebooks narrate varied experiences of "manly love," I read his textual preservation of "specimens" as an incorporative memorial that is intrinsically complex. As the following chapters argue, any understanding of Whitmanian mourning is irrevocably linked not only to the Civil War, but also to the nineteenth-century homosocial experience.

Whitman's notes on the Broadway Hospital Pathological Museum foreshadow his desire to collect "valuable specimens" and record "practical memoranda of the most remarkable cases" that he would soon witness in the military hospitals:

> Here are collected many valuable specimens and practical memoranda of the most remarkable cases that have been treated in the Hospital, for the past fifty years. The curator of the pathological cabinet, &c., is Dr.

> J. J. Hull, who spends much of his time in preparing and preserving for surgical, medical and scientific enlightenment, any marked illustrations of disease, deformity—and also, from time to time, interesting normal specimens of anatomy, &c. These being collected together in the upper story of the building, with the accumulations of past curators and surgeons, and contributions of one kind and another from the medical staff of the Hospital, make a very good museum of its kind. The most horrible and painful liabilities of humanity are exemplified by the memoranda of this cabinet.[185]

The cabinets contain replicas of "the most painful liabilities of humanity," "exactly modeled" from "dead or living subjects." It is unclear from Whitman's account whether he examined actual human specimens, but the poet is clearly fascinated by these anonymous "accumulations" of "disease and deformity." Soon enough, he will see the real thing. As the following chapter will show, the poet's war specimens inhabit the space where mourning and medicine meet each other. With the arrival of so many unidentified casualties and the unprecedented military and medical collaboration to collect and exhibit thousands of "human fragments," the anonymous stranger becomes the disembodied specimen.[186] Both human remnant and anatomical object, the Whitmanian specimen emerges as a melancholically erotic relic that preserves enduring attachments to the "dead young men he might have loved."

2 Specimen Cases
From Hospital to Museum

In the aftermath of the Battle of Fredericksburg, while Whitman reunited with his brother George, the surgeon John H. Brinton scavenged the field hospital's medical waste in search of specimens for the Army Medical Museum, where he served as founding curator. Arriving four days before Christmas, Whitman was immediately confronted by the medical detritus of war. Surgeons operating inside the temporary hospital at Lacy House flung severed limbs from windows bordering the makeshift operating tables.[1] The "rejected members" fell at the base of a catalpa tree below and were eventually buried in mass graves.[2] In a letter to his mother, Whitman described this scene of dismemberment as "one of the first things that met my eyes in camp":[3] "Spent a good part of the day in a large brick mansion, on the banks of the Rappahannock, used as a Hospital since the battle—Seems to have receiv'd only the worst cases. Out doors, at the foot of a tree, within ten yards of the front of the house, I notice a heap of amputated feet, legs, arms, hands & c., a full load for a one-horse cart."

Brinton may have sifted through this same pile of limbs in search of specimens. By the time Whitman entered the scene, the curator was already an established presence in the camp.[4] While both men were compelled to care for the wounded, Brinton's primary mission was "to preserve for the Museum" the "mutilated limbs" that, without his intervention, "were usually buried in heaps."[5] The curator described the nationalistic fervor of his efforts as "infectious": "Many a putrid heap have I dug out of trenches where they had been buried, in the supposition of an everlasting rest, and ghoul-like work I have done, amid surrounding gatherings of wondering surgeons. But all saw that I was in earnest and my example was infectious."[6] Brinton appropriates the viral rhetoric of infection to describe his preservationist mandate. As a curatorial "resurrectionist," he subverted the "everlasting rest" of burial in favor of a public afterlife in the museum.

During the Civil War, the chronic shortage of cadavers available for dissection was suspended. Seemingly overnight, battlefield carnage transformed human bodies from rare commodities, usually obtained illegally, into abundant specimens readily available for the taking. Soldiers' corpses were appropriated by surgeons, embalmers, curators, and photographers; their remains were altered to suit the collector's context. Curiosity about the consequences of battle saw bodies displayed for public consumption long after death and dismemberment. Photographers such as Matthew Brady and Alexander Gardner published wildly popular collections of combat scenes, while thousands viewed the human remnants displayed at the Army Medical Museum.[7]

The spectacle of discarded limbs outside Lacy House endured as a traumatic afterimage in Whitman's memory. In his diary, the poet again reflected on the gruesome "Sight at the Lacy House": "human fragments, cut, bloody, black and blue, swelled and sickening—in the garden, near a row of graves."[8] As a visual legacy, the afterimage returns unbidden, an optical ghost that appears long after the original exposure has ceased. The resonance of this initial encounter foreshadows Whitman's fascination with amputation as a signifier of the wound's erotic vacancy, and his continued reverence toward the war's casualties and their detached parts. Irrevocably altered by these encounters with wounded and dying men, Whitman spent the remaining war years as a constant presence at the bedsides of soldiers in the Washington hospitals, to the detriment of his own health.[9] For the remainder of his life, he sought to preserve their "spiritual character" within his poetry and prose.[10]

Amputation was perhaps the Civil War's most gruesome medical legacy. "The limbs of our friends and countrymen," wrote Oliver Wendell Holmes, "are part of the melancholy harvest which War is sweeping down."[11] That violent reaping severed the extremities of some 60,000 soldiers.[12] Civil War projectiles shattered bones and destroyed surrounding tissue so totally that surgeons were often left with no alternative to amputation.[13] Almost two-thirds of Civil War deaths resulted from infection and disease. The hospital wards were, in their way, as dangerous as the killing fields. Absent of any medical understanding of sepsis and germ theory, the cure was often as deadly as the cause.

Through the entwined narratives of Whitman's *Memoranda during the War* (1875) and the *Personal Memoirs of John H. Brinton, Civil War Surgeon, 1861–1865* (published posthumously in 1914), this chapter explores the shrinking distinction between the human body as an object of mourning and a subject of scientific inquiry during the Civil War. Placing *Memoranda* alongside psychoanalytic frameworks of incorporation, I analyze the hospitality of anonymous mourning and the cultural anxiety resonant around the unburied dead. This collective trauma reflects a dual ambiguity that was deeply troubling to Whitman: the anonymity of the dead and the absence or unlocatability of their graves.

Human Fragments

Memoranda during the War documents the psychosomatic aftermaths of trauma—the embodiment of mourning through the recurrent pain of "old, lingering wounds."[14] Whitman sought to salvage the war's "human fragments"—to textually preserve the "animal purity" of their broken bodies.[15] The word "trauma," drawn from the Greek term for a physical wound, signifies not only an "external body injury" but also a "psychic injury" resulting from "emotional shock, the memory of which is repressed and remains unhealed."[16] Freud emphasized the magnetism of melancholia, which "draw[s] to itself cathectic energies . . . from all directions."[17] The melancholic appropriation of psychic energy parallels Whitman's paradoxical attraction to the hospitals. The wards exerted a mesmeric hold over the poet, revealing to him "the most *magnetic* as well as terrible sight[s]."[18]

Conceptualizing melancholia as a psychic lesion, Freud and Breuer captured the open-ended allure of the unhealing wound. They described the "memory of trauma" as an agent provocateur that operates as a "foreign body" lodged within the mind, continuing to wreak havoc "long after its forcible entrance."[19] This conception of trauma as both open wound and embedded shrapnel correlates with the Army Medical Museum's mandate to collect artifacts demonstrating both unusual wounds and the projectiles removed from them: "Medical Officers are directed diligently to collect specimens of morbid anatomy . . . together with projectiles and foreign bodies removed . . . as may prove of interest in the study of military medicine and surgery."[20] While the overt goal of the museum was to preserve artifacts of "lasting scientific interest," it operated within the "pathological public sphere" as a phenomenon that Mark Seltzer has described as an "atrocity exhibition."[21] This medical spectacle catalyzed "a fascination with the shock contact between bodies and technologies," appealing to a public captivated by the human cost of recent innovations in weaponized warfare.[22]

"As soon as the Museum was fairly established," Brinton wrote, "it began to attract attention. The public came to see the bones, attracted by a new sensation."[23]

Brinton's dual use of the verb "attract" to describe the gravitational pull of this "new sensation" speaks volumes about the museum's popularity. Visitors were drawn "to see the bones," their gaze held by specimens originating from the convalescent tableaus that captivated Whitman. Unlike Whitman's devotion to living soldiers, the Army Medical Museum evokes a quarantined, posthumous voyeurism. These medical specimens are enclosed within cases that prevent them from contaminating the living. The exhibition is not only sensational, it is also, Brinton emphasizes, *new*. The Medical Museum operates as an early example of Seltzer's "wound culture," in which "the very notion of sociality is bound to the excitations of the torn and opened body, the torn and exposed individual, as public spectacle."[24] I interrogate the Army Medical Museum as a cabinet of anatomical curiosities that exemplifies postbellum America's preservation compulsion. Brinton's memoirs enact a collective suturing function. The curator inscribes a medical mandate onto the war's disparate fragments. The Army Medical Museum enshrines the *sources* of trauma—extracted projectiles—alongside remains of the human bodies they shattered. Brinton merges foreign, military, and human bodies to create a Unionist narrative of the war's medical legacy. Yet, as we shall see, his is also an account of *pathological detachment*, in which the human body is repeatedly subjected to anatomical invasions and exhibitions, often without consent and at times in direct conflict with the soldier's expressed wishes.

Reflecting on the influx of the wounded into the Washington hospitals, Whitman casts these men as emblems of democracy:

> In my recollections of that period, and through its varied, multitudinous oceans and murky whirls, appear the central resolution and sternness of the bulk of the average American People, animated in Soul by a definite purpose, though sweeping and fluid as some great storm . . . emblemized in thousands of specimens of first-class Heroism. . . . To me the main interest of the War, I found (and still, on recollection, find) in those specimens, and in the ambulance, the Hospital, and even the dead on the field. To me, the points illustrating the latent Personal Character and eligibilities of These States, in the two or three millions of American . . . men, North and South, embodied in the armies—and especially the one-third or one-fourth of their number, stricken by wounds or disease at some time in the course of the contest—were of more significance even than the Political interests involved.[25]

Through the "murky whirls" of memory, Whitman's specimens emerge as dual archetypes of the "average American" "Soul" and the exceptional "Heroism" "embodied in the armies." As an incarnation of the fractured Union and the fragmented body, they represent "countless" fallen soldiers.[26] Whitman remained haunted by the anonymous dead, those who were "inhumed by the strangers," lay nameless in mass graves, or remained where they fell, "unburied and unknown."[27] Their deaths eclipse all other consequences, as though the entire war was waged upon the specimen as an "emblem" of the democratic body: "Not Northern soldiers only . . . many a Southern face and form, pale, emaciated, with that strange tie of confidence and love between us, welded by sickness, pain of wounds."[28] This "strange tie" of mutual suffering "weld[s]" together "representatives" from across the divided states. The whole (union) is signified by one of its parts—the genus is named for the species. The specimen embodies the synecdoche of the stranger: an anonymous other capable of absorbing the diversity of war casualties.[29]

In the Army Medical Museum, the specimen is a detached part of a specific body, preserved from decay and bearing no visibly discernible identifying markers. Divorced from its "original possessor," the museum specimen attains a form of corporeal immortality; it becomes, in Brinton's words, a relic of "lasting scientific interest."[30] Examining the intersection between these two methods of collection (medical and poetic), I find that Whitman radically departs from the medical museum model of dehumanization. For Whitman, the war's "human fragments" epitomize the enduring sublimity of the body. Traces of its "spiritual character" endure beyond death and dismemberment.[31] I analyze the Whitmanian specimen as material and psychic remains of the dying soldier: a synecdochic figure that facilitates the symbolic burial of countless inaccessible bodies.

The Civil War ruptured the panoramic inclusiveness of *Leaves of Grass*, altering Whitman's understanding of nationalism, and the function of his text within the cultural landscape. In the preface to "Good-Bye My Fancy" (1891–92), Whitman described the war's impact on his lifelong project:

> Those hot, sad, wrenching times . . . the wounded, suffering, dying—the exhausting, sweating summers, marches, battles, carnage—those trenches hurriedly heap'd by the corpse-thousands, mainly unknown—Will the America of the future—will this vast rich Union ever realize what itself cost, back there after all?—those hecatombs of battle-

> deaths—Those times of which, O far-off reader, this whole book is indeed finally but a reminiscent memorial from thence by me to you?[32]

The "unending" task of mourning the war's "infinite dead" radically extended Whitman's open-ended approach to textuality. The postbellum structure of *Leaves of Grass* echoes the psychic resurrection of trauma—an event so arresting that it is never "fully known and is therefore not available to consciousness until it imposes itself, repeatedly, in the nightmares and repetitive actions of the survivor."[33] Further expanding Whitman's lifelong revision process, the war poems were repeatedly clustered, altered, or expelled in the author's quest for an elusive body of work that could function as a "reminiscent memorial." This "compulsion to repeat" through the evolution of poetic work speaks to the fundamental dissonance of traumatic experience: the possibility of pleasure through suffering.[34]

Throughout his hospital visitations, Whitman recorded soldiers' case histories: "I kept little note-books for impromptu jottings in pencil. . . . In these I brief'd cases, persons, sights, occurrences in camp, by the bedside, and not seldom by the corpses of the dead." Whitman eventually incorporated revised versions of the notebooks into *Memoranda during the War* (1875–76). *Memoranda* was later absorbed, almost entirely, into *Specimen Days* (1882). Yet elements of the original, "blood-stain[ed]" pages were withheld; these phantoms and their "associations" were accessible only to Whitman. As Buinicki observes, "Whitman presents his notes primarily as supplements of memory, before underlining how close he has attempted to remain to the original experience. . . . What were originally designed to refresh Whitman's memory while on the spot have become the memories themselves."[35] The notebooks resist narrative coherence, forming instead a collage of details that might otherwise be forgotten. Their pages recall the last words of soldiers and the resonant silence of the dead, whose bodies would become the war's most "eloquent bequest."[36] Whitman describes the notebooks as blood relics of history:

> Forming a special history of those years, for myself alone, full of associations never to be possibly said or sung. I wish I could convey to the reader the associations that attach to these soil'd and creas'd little livraisons, each composed of a sheet or two of paper, folded small to carry in the pocket, and fasten'd with a pin. I leave them just as I threw them by during the War, blotch'd here and there with more than one blood-stain, hurriedly written.[37]

This text holds an incantatory power because it retains the sense of partiality in which it was composed.[38] Whitman's journals are emotive conduits into the traumatic urgency of the past: "Vivid as life, they recall and identify the long Hospital Wards . . . the convulsive memories (let but a word, a broken sentence, serve to recall them)."[39] Like the flashbacks they invite, the books are both "convulsive" and fractured: their words are "broken," their pages linked only by a pin. The notebooks form a "special history" known only to the poet, containing "associations" that defy representation, "never possibly said or sung."[40]

In the hospitals' convalescent spaces, through acts of hospitality to dead and dying men, Whitman experienced profound loss, but also heightened attachments to liminal beings, those on the threshold between life and death.[41] These encounters were all the more lasting for their ephemerality, and all the more pleasurable for their juxtaposition with intense pain: "I believe no men ever loved each other as I and some of these poor wounded, sick and dying men love each other."[42] In the absence of burial for many fallen soldiers, Whitman devised an alternative method for mourning the dead: the specimen is preserved within the text, resurrected whenever its pages are opened:

> I can never turn their tiny leaves . . . without the actual army sights and hot emotions of the time rushing like a river in full tide through me. Each line, each scrawl, each memorandum, has its history. . . . Out of them arise active and breathing forms. They summon up, even in this silent and vacant room as I write, not only the sinewy regiments and brigades . . . but the countless phantoms of those who fell and were hastily buried by wholesale in the battle-pits, or whose dust and bones have since been removed to the National Cemeteries of the land.[43]

Memoranda narrates the harrowing effects of Whitman's encounters with his hospital notebooks. Both writing and reading are haunted acts. The old text (the notebooks) and the new (the revised *Memoranda*) both "summon" the dead. The poet allows his readers to glimpse the "breathing forms" conjured by tactile contact with the bloodstained traces of his lost soldiers. Yet we are not given unfettered access. We require the narration of a poet-medium, one who was actually present "amid those scenes."[44] Whitman's response to these "tiny leaves" goes beyond the provocation of a traumatic flashback. "Living" beings "arise" from the pages, bringing with them the "actual" "sights" and "emotions of the time." The book acts as a psychic crypt for a collective haunting, home to "countless phantoms" whose "dust and bones" were relocated to the national cemeteries.

The exhumation and reburial of their bodies reflects the diasporic afterlives of Civil War specimens.

The impossibility of literary representation mirrors the psychological manifestation of trauma, an experience so sudden and shattering that it cannot initially be fully integrated.[45] Whitman's revisions function as elements of a repetition compulsion, allowing him to return, "in dream's projections," to the hospital corridors.[46] Each incarnation of the war text is an act of incorporative mourning. The bloodstained original is absorbed into the latest work, slightly altered with each retelling. Whitman's halting prose mirrors the elusive magnetism of trauma. It is not easily integrated into a linear narrative. Its ghosts arrive without warning, arising from the pages, "vivid as life."[47] Even the act of writing evokes violence, as does the method of creating the book: the words are "scratch'd," the pages sutured with a pin.[48] As a discourse on the psychosomatic resonance of trauma, *Memoranda during the War* occupies a position of startling contemporary relevance, not only to literary representations of history, but also to issues of legacy, mourning, and the inherent unknowability of the casualties of war.

Ghoul-like Work

John Hill Brinton volunteered with the surgical brigade in August 1861. Within a year, he was reassigned to the Surgeon General's Office, where his primary duty was "to prepare the Surgical History of the Rebellion."[49] On August 1, 1862, Brinton was directed by Surgeon General William A. Hammond to establish a museum devoted to "specimens of morbid anatomy."[50] This was the beginning of a collection that would incorporate thousands of Civil War remains, many of which are still on display at the National Museum of Health and Medicine.

Brinton soon found himself entrenched in a "ghoul-like duty," responsible for the procurement of anatomical artifacts "portraying the results of wounds, operations, or peculiar amputations."[51] He traveled to hospitals and battlefields in search of specimens, sometimes exhuming bodies from their graves in order to obtain the desired part. He instructed surgeons on the best methods of preservation: how to pack limbs in kegs of alcohol and arrange their safe transportation to Washington. Brinton regarded the curatorship as his greatest professional achievement, stating simply: "my whole heart was in the Museum . . . by it the results of the surgery of this war would be preserved for all time."[52] Reflecting on his war service decades later, the curator nostalgically inserted a part of himself into the anatomical collection. His heart is incorporated within the museum that is devoted to ensuring the war's surgical legacy.

Brinton's rhetoric reflects a culture of preservation that began during the war and flourished in the aftermath of Lincoln's assassination. The museum infused dissected matter with national significance, in an effort to counter the symbolic dismemberment of the Union. As detailed in chapter 4, the extended public display of Abraham Lincoln's body was made possible by recent innovations in embalming, practiced on the bodies of unknown soldiers.[53] Lincoln eventually joined the ranks of specimens whose case histories were detailed in the six-volume *Medical and Surgical History of the War of the Rebellion* (1870–83). The museum was relocated to Ford's Theatre in 1866, securing Lincoln's status as America's most sacred democratic specimen.[54]

Brinton's devotion to specimens preceded the war's onset. Prior to departing for military service, his "final preparations" were to "care for the preservation of my anatomical possessions."[55] The surgeon's assertion of ownership over these human relics recalls the lingering philosophical quandary of antebellum America: how can one individual own the body of another? As chapter 1 established, this ethical paradigm connected live bodies with dead ones, linking arguments against medical body snatching with abolitionism. Corporeal possession is an immutable element in the narrative of American nationalism: the Union wielded the power to legislate the appropriation of certain bodies (living or dead, whole or in part).

In the opening pages of his memoir, Brinton frames the crisis of the Union in terms of preservation: "the problem was how to preserve the unity and majesty of the nation, and how soonest to trample out the doctrines of disintegration and 'secession.'"[56] The surgeon's preservation compulsion entails not only an anatomical agenda, but also a patriotic impulse. Brinton created a coherent medical narrative of war, reflecting a nation that conserved the "rejected members" of its citizens.[57] The Army Medical Museum echoed the Unionist ideology that decay of the wounded democratic body could be arrested, that amputation of the secessionist states was not the only option.

Despite his horror at the conditions inside Lacy House, Whitman resisted the temptation to sentimentalize what was in effect the necessary disposal of medical waste in a war triage setting. He recognized the chaos in which the surgeons worked: "The large mansion is quite crowded, upstairs and down, everything impromptu, no system, all bad enough, but I have no doubt the best that can be done; all the wounds pretty bad, some frightful, the men in their old clothes, unclean and bloody."[58] Reflecting on the perils of "ignorant physicians," Brinton recalled a similarly bloody environment at Fort Donelson. He was far more scathing in his indictment of unsanitary conditions. After hearing accounts of a "great surgeon" operating from a hospital established in a "little country house," he set out to observe the man at work: "I found bloodstained footmarks on the crooked stairs, and in the second story room . . . amputated arms and legs seemed almost to litter the floor; beneath the operating table was a pool of blood, the operator was smeared with it and the surroundings were ghastly beyond all limits of surgical propriety."[59] Even in the era before pathogenic medicine, Brinton was instinctually repelled by the absence of hygiene. Medical practices that existed within the boundaries of "surgical propriety" were often equally ghastly. Both armies faced a shortage of trained doctors, and resorted to the employment of "contract physicians" who were sometimes woefully inexperienced.[60] Brinton, who had served as president of the Medical Examining Board, decried the presence of "imposters" within the surgical brigade, "long out of practice, if indeed they ever had any."[61] Brinton's fury toward inept doctors was echoed on both sides of the Mason-Dixon line. The Confederate surgeon Julian John Chisholm deplored the threat of the scalpel in overzealous hands: "The limbs of soldiers were in as much danger from the ardor of young surgeons as from the missiles of the enemy." In the Confederate textbook for military surgeons that he authored, Chisholm alleged that opportunistic surgeons were practicing their craft on soldiers: "Among a certain class of surgeons . . . amputations have often been performed when limbs could have been saved, and the amputating knife has often been brandished by inexperienced surgeons, over simple flesh wounds."[62] Echoing antebellum debates surrounding the ethics of medical body snatching, Chisholm accused unethical surgeons of conducting human vivisection.

As casualties overflowed Lacy House, makeshift cots were fashioned in the surrounding garden. Here Whitman began the bedside ministrations that would consume him for the remainder of the war:

> The wounded lying on the ground, lucky if their blankets are spread on layers of pine or hemlock twigs, or small leaves . . . . I go around from one case to another. I do not see that I do much good to these wounded and dying; but I cannot leave them. Once in a while some youngster holds on to me convulsively, and I do what I can for him; at any rate, stop with him and sit near him for hours, if he wishes it.[63]

This soldier's "convulsive" embrace demonstrates the permanence of Whitman's ties to the men he attended.[64] However immaterial he perceives his presence to be, he is incapable of leaving them behind. Medically speaking, convulsions are involuntary muscle contractions that induce spasms or seizures. To convulse is to unite by tearing, to violently pull disparate entities together. Like melancholia, convulsiveness is magnetic, drawing energy ever inward.[65] The wound, too, is a convulsive host. As it heals, a wound invites new tissue growth. If it festers, it attracts bacteria and decay. *Leaves of Grass* mirrors this paradoxical hospitality. Unlike Freud's rapacious psychic wound, Whitman's embodied mourning was symbiotic, though not without a cost to its host. The poet's health steadily declined in the postwar years; he never recovered from the mental and physical exhaustion of the hospitals. Freudian melancholia leaches energy, "emptying the ego until it is totally impoverished."[66] Not so for Whitman. The war may have destroyed his health, but it gave him something far greater in return. Toward the close of his life, Whitman confessed to Horace Traubel that, while his hospital work ravaged him physically, it was also the primary source of his "consummated book":

> I had to pay much for what I got but what I got made what I paid for it much as it was seem cheap. I had to give up my health for it—my body—the vitality of my physical self. . . . What did I get? Well—I got the boys, for one thing: the boys: thousands of them: they were, they are, they will be mine. I gave myself for them: myself: I got the boys: then I got *Leaves of Grass*: but for this I would never have had *Leaves of Grass*—the consummated book (the last confirming word): I got that: the boys, the Leaves: I got them.[67]

From the leaves that cushioned soldiers' camp beds to the "last confirming word[s]" of the "deathbed edition," *Leaves of Grass* was entwined with the war's broken bodies. Whitman's remarks reflect an economy of anatomical and textual exchange. In order to claim the boys who would "consummate" his book, the author must sacrifice himself. In exchange for relinquishing his own physical "vitality," the poet assumes possession of his "boys." Yet it is not only the soldiers who are preserved by this logic; the "deathbed edition" (and by extension the entire postwar canon) would not have been possible without Whitman's war experiences. *Memoranda during the War* charts Whitman's vast alteration through the hospitality of merging with another's wound. From his earliest war entries, Whitman insists that the body need not be whole, or even alive, in order to be adored.[68]

1. Alexander Gardner and John Reekie, *A Burial Party, Cold Harbor, Virginia*, 1865. From Gardner's *Photographic Sketchbook of the War*. Library of Congress. Prints and Photographs Division.

Specimens of Unworldliness

When he departed Falmouth on December 28, 1862, Whitman accompanied a convoy of wounded soldiers bound for the Washington hospitals. They travelled by rail to Aquia Creek, Virginia, where they transferred to a government steamship. Whitman recorded the casualties of their arduous passage: "On the boat I had my hands full. One poor fellow died going up."[69] Several days earlier, Brinton and his assistant curator, William Moss, made arrangements for the transportation of their museum relics, as the wounded were evacuated from the battlefield.[70] They were ambushed while smuggling "an immense number of surgical specimens . . . on the backs of one or two very black negroes."[71] After surviving Confederate gunfire with their lives and specimens intact, Moss exclaimed, "What a blessed escape, for what a wretched ending it would have been to one's life, to have been swept into the river on an ignominious retreat, holding onto a bag of bones."[72] Brinton's account of their narrow escape demonstrates the totality of the Union's appropriation according to military and racial hierarchies. He ignores the earlier admission that these specimens were not actually carried by the surgeons themselves. African American laborers transported the remains across enemy lines. (See figure 1.) Lamenting how "wretched" it would have been to die in retreat, clutching his "bag of bones," Moss assumes possession of the "black" hands that actually held the remains, as though they were, by extension, his own. In another rhetorical twist, he adopts the wrong body part—having initially specified that the bones were carried on the men's backs, a position that links them with slavery and animality. Even the number of black bodies is uncertain. When assuming credit for the labor of specimen transportation, Brinton rhetorically transfers the objects from the backs of black men into the hands of white surgeons.[73]

While Brinton expanded the museum's collections, Whitman established residency in the nation's capital as a "self-appointed Soldier's missionary."[74] The poet devoted his considerable energies to the hospitals, tending, by his own estimation, "80,000 to 100,000 of the wounded and sick, as sustainer of spirit and body."[75] Whitman's letters establish a ritualistic pattern: he visited the hospitals twice a day, usually from noon until four, then returned in the evening, often lingering all night with "dear or critical cases."[76] From October 1861 until March 1863, the Patent Office relinquished the second-floor gallery for the care of wounded soldiers. Whitman had long admired the architecture of its Doric façade, describing it as "that noblest of Washington buildings."[77] Throughout the nineteenth century the galleries displayed models submitted by inventors alongside their patent applications, and provided exhibition space for artifacts of national significance.[78] Commissioner Henry Ellsworth described the National Gallery housed within the Patent Office as "a perpetual exhibition of progress and improvement. . . . Here the most beautiful specimens of the genius and industry of the nation will be found."[79] Whitman was mesmerized by the "immense apartments filled with high and ponderous glass cases, crowded with models in miniature of every kind of utensil, machine or invention it ever entered the mind of man to conceive; and with curiosities."[80] (See figure 2.)

A decade before the Civil War, the galleries housed a posthumous menagerie of natural history specimens. In 1836 Congress funded the U.S. Exploring Expedition (commonly known as the Wilkes Expedition after its commander, Charles Wilkes) to survey the Pacific Ocean and South Sea Islands.[81] A corps of scientists—including naturalists, botanists, and taxidermists—collected specimens throughout the voyage. They gathered thousands of zoological and botanical artifacts, including 2,000 previously unidentified species, which founded the Smithsonian's natural history holdings.[82]

During the war, the Model Room contained an even stranger collection. The most severe casualties from the Second Battle of Bull Run, Antietam, and

2. Old patent office, model room, 1861–65. Library of Congress. Prints and Photographs Division, Brady-Handy Collection.

Fredericksburg were brought here. The sight of hospital cots scattered between the illuminated display cases created an eerie spectacle:

> It was a strange, solemn and, with all its features of suffering and death, a sort of fascinating sight. . . . Between these cases are lateral openings . . . and in these were placed the sick . . . Many of them were very bad cases, wounds and amputations. . . . It was, indeed a curious scene at night, when all lit up. The glass cases, the beds, the forms lying there,

the gallery above, and the marble pavement under foot—the suffering, and the fortitude to bear it in various degree . . . sometimes a poor fellow dying, with emaciated face and glassy eye, the nurse by his side, the doctor also there, but no friend, no relative—such were the sights but lately in the Patent Office.[83]

This "curious scene," perhaps more than any other, haunted Whitman's hospital prose. Traces of these human specimens recur as ghostly imprints scattered throughout the Whitman canon. Their bodies recall the industrial models and zoological specimens that surrounded them, and preceded their occupation of this space. Their convalescence within the galleries foreshadows their potential afterlives, dismembered and stripped of flesh, in the Army Medical Museum. These soldiers died among strangers, with "no friend, no relative" to witness their final hours. Anonymity is central to their spectral magnetism. Throughout Whitman's war poetry and prose, the unknown soldier is an enduring figure of collective grief: "Unnamed, unknown, remain, and still remain, the bravest soldiers."[84] How can we mourn the dead when "their very names are lost," when their bodies remain unburied?[85] This is a central question of *Memoranda, Drum-Taps*, and postbellum editions of *Leaves of Grass,* which the figure of the specimen attempts to reconcile. As a representative body capable of merging with others, who died similar deaths in similar places, the specimen allows the act of mourning to be unbroken by the limits of selfhood and otherness, known and unknown.

Within the corridors of the model room, Whitman's soldiers exist as uncanny doubles, haunted by past and future specimens. Initially, they are framed by the cabinets of curiosities that decorated the galleries. Two years later, at Lincoln's second inaugural ball, Whitman saw the scenes superimposed: revelers danced between unseen and uninvited guests, the ghosts of lost soldiers: "To-night, beautiful women, perfumes, the violins' sweetness, the polka and the waltz; then the amputation, the blue face, the groan, the glassy eye of the dying; the clotted rag, the odor of wounds and blood, and many a mother's son amid strangers, passing away unattended there."[86] These resurrected patients invade the present with phantasmal sensations. The ladies' perfumes are obscured by the "odor of wounds and blood"; the "violins' sweetness" is drowned out by the "groan[s]" of the dying. The inauguration is superseded by the illumined bodies reflected in the glass cabinets, their suffering rendering them somehow transcendent.

As a hallucinatory ghost, the afterimage mirrors the phantom pains experienced by many of the amputees Whitman attended. While the afterimage

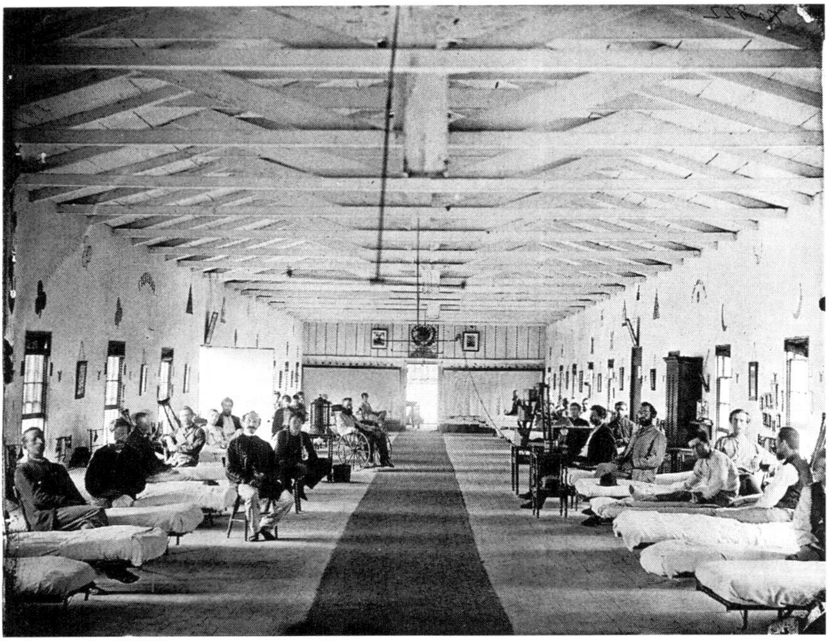

3. Patients in Ward K, Armory Square Hospital, Washington DC, 1865. Library of Congress. Prints and Photographs Division.

functions as the visual trace of a trauma one continues to see, the phantom limb embodies the absence of an entity one continues to feel. As chapter 3 will show, Whitman's sensory haunting demonstrates the physicality of mourning: it is inscribed upon the body. The past returns as physical sensation: sight, smell, and sound. The poet's specimen ghosts surround and consume him. They are not confined to the interiority of the mind.

After the Patent Office wards closed in March 1863, Whitman spent most of his time at Armory Square on the Mall. (See figure 3.) As he explained to his mother, "I devote myself much to Armory Square Hospital because it contains by far the worst cases, most repulsive wounds, has the most suffering & most need of consolation."[87] The following archival fragment invites the reader to see the wards through Whitman's eyes:

> Would you not like to see for yourself, Dear reader, some special ones of the cases among the hundreds I have met? Enter with me this long ward, look down its rows of cots, with their occupants stretching away each side. With the wide open aisle in the middle. Every one of these cots has

> its history—every case is a tragic poem, an epic, a romance, a pensive and absorbing book, if it were only written.[88]

Each "case" issues the invitation of a blank page. Collectively, they form a "pensive and absorbing book"—a spectral text that exists, as yet, only in the mind of its author and the bodies of its subjects. This representation of the convalescent body as an "absorbing book" portrays Whitman's war texts as sites of incorporation for the war's unwritten histories and unburied corpses. The book becomes a textual crypt that houses the specters of lost soldiers, allowing author and reader continued access to their afterlives. Whitman attests to the literary capacity for resurrection—the ability of words, however fragmented, to haunt.

In Blooming Flesh

The Army Medical Museum's first specimens were sourced from the same wards that Whitman frequented. Brinton recalled the limbs' transformation from "human fragments" into pathological artifacts: "I obtained amputated arms and legs from the Washington hospitals . . . these [were] cleaned, prepared and mounted, and . . . made their first appearance on top of my desk."[89] The curator's account of the specimen's journey from human appendage to medical waste to museum relic illuminates his memoir as a narrative of corporeal transference:

> Any account of the Museum would be incomplete without some description of how. . . they passed from their original possessors to the Museum. . . . The bones of a part removed would usually be partially cleaned, and then with a wooden tag and carved number attached, would be packed away in a keg containing alcohol . . . sent to Washington and turned over to the Army Medical Museum . . . so that they could take their place on the shelves.[90]

Brinton's preservation compulsion was not only physical, but also textual. He meticulously archived the histories of his "objects": "The memoranda . . . [were] forwarded to the Surgeon-General's Office, . . . entered in the books of *Histories of Specimens*, preserved in the Museum."[91] The aesthetics of display were equally paramount. The curator designed glass cabinets modeled after those in his home office: "I was enabled . . . to fit up good cases for the rapidly growing collection. The doors locked with bronze hands, which slid bolts at top and bottom."[92] The latches are gilded reminders of the human bodies from which the enclosed remains were sourced. These bones are further dehumanized by Brinton's invocation of human anatomy to describe the locks, but not the specimens themselves. This anatomical amnesia recalls another pair of hands appropriated by the museum: the black hands that smuggled specimens across enemy lines at Fredericksburg.[93]

The curator's "search for specimens" led to many "strange scenes" of exhumation. "One such case" concerned the acquisition of "a remarkable injury of the lower extremity."[94] Brinton's efforts to secure the specimen were initially thwarted: "the man had died with the limb on and had been carefully buried by his comrades," who were determined to protect the corpse from disturbance. Unwavering in his determination, Brinton visited the "mess mates, explained my object," and "dwelt upon the glory of a patriot having *part* of his body at least under the special guard of his country."

Brinton not only convinced the soldiers to agree to the disinterment, but to carry out the act themselves: "the comrades of the dead soldier solemnly decided that I should have the bone for the good of the country, and in a body they marched out and dug up the body." The curator persuaded the fallen soldier's comrades with the Unionist lure of synecdochic glory: "a patriot" giving up "a *part* of his body . . . for the good of the country."[95] This logic demonstrates the uncanny duality of military bodies: the soldiers form a patriotic body that exhumes a corporeal one. The specimen assumes a life of its own, independent of the vanishing agency of its "original possessor."[96] In addition to antebellum arguments that medical advancement warranted the appropriation of indigent bodies without the consent of the decedents or their survivors, resurrection now enjoyed a patriotic mandate. Soldiers who had already donated their living bodies "for the good of their country" could attain posthumous or post-amputation honor by relinquishing their severed limbs to the museum.

Brinton recorded the common drive of many amputees to revisit their lost parts: "Officers and soldiers who had lost a limb by amputation would often come to look up its resting place."[97] One account of a soldier observing his specimen recurs several times throughout the curator's published materials:

> On one occasion a man from the ranks demanded the return of a limb . . . He was informed that the member in question could not be given up. "But it is mine," he said, "part of myself," earnestly enforcing his claim, and his demand to the lay mind seemed reasonable. Yet to surrender a specimen was very much like yielding a principle. "Stop," said the quick-witted young assistant curator to the claimant, "for how

long have you enlisted?" "For three years or the war." "Then, replied the official, "the contract is not yet terminated, come back at the end of the war or at the termination of your three years' service and you can have your bone. In the meantime one detachment of you is stationed in this Museum on government duty, the other wherever you may be ordered."[98]

This soldier resists the museum's doctrine of military possession, seeking to reclaim the limb as "his *own* property."[99] The curator insists that his enlistment mandates governmental ownership over his entire body. One "detachment" remains "on duty" as a medical spectacle stationed within the museum. The Union retains whatever "part" of the military body remains useful, irrespective of the psychological consequences.

The Washington author Mary Clemmer Ames included a chapter on "The Army Medical Museum—Its Curiosities and Wonders" in her memoir *Ten Years in Washington: Life and Scenes in the National Capital, as a Woman Sees Them* (1873). Ames captures the museum's ornate aesthetic, describing a cranial specimen, "as white as crystal; mounted on gold, tiny blue and crimson threads of silk trace from chin to head-top the entire nerve system. It is a work of exquisite art as well as of science, and in no sense repulsive."[100] Despite her awe at this gilded skeletal articulation, Ames is not blind to the collection's harrowing resonance: "to the unscientific mind, especially to one still aching with the memories of war, it must ever remain a museum of horrors."[101] Nevertheless, she catalogs the military specimens with the reverence of a disciple examining the remains of martyrs: "Its many bones, which never ached, and which have survived their painful sheaths of mortal flesh, all cool and clean, and rehung on golden threads, are not unpleasant to behold."[102] For Ames, the specimen-relic has transcended human suffering. Once their "sheaths of mortal flesh" are cast aside, the soldiers' bones have "never ached" again.

These sanctified military remains came to reside in stark contrast to another collection of human artifacts: skeletal relics of social Darwinism. Ames engages her readers in a virtual tour of the museum, lingering on the models of craniometry: "At the right of the main entrance, stands the Craniological Cabinet. It contains a thousand or more specimens of the craniae of different human races. Beside the skull of the Caucasian, we see that of the African, each of the highest order of its kind."[103] Classified hierarchically by race, these skulls were surrounded by specimens from the natural world, enhancing their perceived animality.[104] This categorical system adheres to the museum's narrative of nationalistic possession, dividing bodies into those that invited either patriotic worship or morbid curiosity. Ames descends into racist dehumanization as she describes Native American casualties. In stark contrast to the soldiers' canonized skeletons, the "Skull of Little Bear's Squaw" warrants only a cursory inventory of wounds: "Perforated by seven bullet holes. Killed in Wyoming Territory."[105] The Sioux infant, a "little papoose mummy," invites a degree of maternal compassion: "This melancholy child also wears a white necklace, and was found *buried* in a tree . . . according to the custom of the tribe."[106] The infant was exhumed from his arboreal grave, a funeral ritual rendered as curiously exotic as his "profusely ornamented" burial costume. Ames's sympathy for the "melancholy child" is contingent upon his untimely demise: "it did not live to slay our brethren, so we are sorry as we look on it."[107] An adult male Sioux, figured as a threat to her race, would be unworthy of sentiment.

Brinton was not responsible for the acquisition or display of these remains. They arrived at the museum after his curatorship ended in 1864. Yet their exhibition is an extension of the exhumation compulsion that began with medical body snatching and continued with the preserved remains of Civil War soldiers. While government officials compiled war statistics and arranged for the repatriation of Union corpses to national cemeteries, army surgeons in the West exhumed the remains of hundreds of Native Americans. As Ann Fabian explains, the reburial of Civil War soldiers and the theft of Native American remains "drew on the same personnel and employed the same ideas and the same tools, albeit to very different ends for communities of survivors."[108] While military casualties were rescued from mass graves and reinterred in national cemeteries, indigenous bodies were removed from ritualized burial grounds, destined for curiosity cabinets. The museum's uncanny appropriation of corpses (both familiar and other) extends antebellum discourses on anatomy, when dissection was limited to the bodies of those deemed socially expendable. In the aftermath of war's inclusive carnage, all of the remains within the museum were subjected to the anatomist's scalpel. Yet they continued to be classified according to those deemed worthy or unworthy of national mourning: either soldiers or strangers.

The Marrow of the Tragedy

After Lincoln's assassination, Ford's Theatre was remodeled to install "long rows of glass cases, in which [were] exhibited to the glances of the curious the prepared specimens of anatomy."[109] The renovated museum opened in its new location on April 13, 1867.[110] By the year's end, more than 6,000 visitors had examined the collection.[111] Dr. John Eric Erichsen, a visiting fellow from London's University College Hospital, described the Army Medical Museum as "occupy[ing] a building that has a melancholy interest."[112] Ames reflected on the sacred status conferred upon the scene of Lincoln's murder:

> It was well that [Ford's Theatre] should be consecrated to a national purpose. None could be more fit than to make it the repository of the Pathological and Surgical results of the war. In two years from the day of the tragedy, its doors were opened to the people, to come in and behold what war, disease, death and human skill hath wrought.[113]

To commemorate the anniversary of Lincoln's death, these military specimens were transferred to hallowed ground, "consecrated" by the president's blood. Katherine Kinney has described the Army Medical Museum as a tourist destination designed to capitalize on violence: "History here is literally the abstraction of bodies into processes of power, production, and consumption. Carnage and assassination are comprehended and displayed as a tourist attraction."[114] Ames's description of the museum as showcasing "war, disease, death and human skill" offers an incongruous yet accurate portrayal of Brinton's meticulously preserved and lavishly displayed collection, juxtaposing shattered bones with the weapons that destroyed them. Given Whitman's fascination with war specimens and Lincoln's death scene, he may well have been among the throng of visitors. The following draft is, in all likelihood, a response to the Army Medical Museum:

> The mouldering bones and dry skeleton or parts of the skeleton are all that is presented as Past History. But that is not Past History. The Past! The peoples of a hundred or a thousand, or ten or twenty thousand—yea fifty or a hundred thousand years ago, they too lived in blooming flesh, with sparking eyes and speaking lips, knew love, ambition, war!, perhaps even science the same as we do now.[115]

This fragment captures Whitman's fascination with the instability of history. In *Specimen Days*, the poet insists that "the real war will never get in the books"—or in this case, onto the museum's shelves. These "dry bones" cannot resurrect the "blooming flesh" of lost soldiers. Even poetry can only capture fragmentary "glimpses" of suffering and tenderness: "the profuse beauty of the young men's hair damp with their spotted blood."[116] Buinicki articulates the paradox of Whitman's war writing—the inevitable limits of language that constrict his ability to incarnate the "blooming flesh" of those lost to history:

> For Whitman, the written word alone is never enough, offering as it does only "scraps and distortions" of events. . . . Whitman is left to describe his attempts to transcend the signifying nature of language and to provide a more embodied connection between the reader and the text . . . the goal is not simply an intimate embrace, but the preservation of memory in the face of the calcifying grasp of history.[117]

Despite his assertion that the traumas of war remain untranslatable, Whitman endeavored to construct a text that could offer, at least, a glancing approximation. He hoped that *Memoranda* would "furnish a few stray glimpses into . . . those lurid interiors . . . never to be fully convey'd to the future."[118] Whitman's rhetoric reveals a resurrectionist desire to unearth the war's "untold" history, alongside a paradoxical sense that certain aspects of that history must remain "buried":

> The marrow of the tragedy concentrated in those Army Hospitals . . . those forming the Untold and Unwritten history of the war—infinitely greater (like Life's) than the few scraps and distortions that are ever told or written. Think how much, and of importance, will be—how much civic and military, has already been—buried in the grave, in eternal darkness.[119]

Given the publicity of the Army Medical Museum and its catalog, Whitman's anatomical metaphor of hospitals as "marrow" alludes to *The Medical and Surgical History of the War of the Rebellion* (1870–83). These six volumes contain summaries of over 6,000 specimens, alongside casualty statistics, etchings, engravings, and photographs. Brinton worked exhaustively on the *Surgical History* until he was relieved from the Surgeon General's Office in 1864.[120] Similar in scope to nineteenth-century analyses of natural history collections, the compilers approached war as a human experiment that allowed researchers to accumulate unparalleled quantities of medical data.[121] As a counterpoint to this statistical analysis, Whitman describes the "marrow of the tragedy" as "buried" and "unwritten"—impossible to exhume either medically or poetically.

A letter from William O'Connor to Whitman (October 2, 1884) reveals that many years after the war, when the poet was residing at Camden in declining health, he remained interested in the contents of the Army Medical Museum and desired access to its archives:

> I have been over today to the Surgeon General's office to see about data for you. . . . I am afraid that the quest will be fruitless. The only matter they have is the Medical and Surgical History of the War, now in process of publication, what you want— i.e. hospital matter—will be in the third volume . . . I will go down tomorrow to the Medical Museum . . . talk with Dr. Wild, the librarian, and see if he can give me anything. I fear it is unlikely—the publications being inchoate.[122]

O'Connor's description of the archival material as "inchoate" reflects the ephemerality of war casualties. The quest to compile mortality figures had become a national obsession following the war.[123] Whitman echoes this cultural anxiety surrounding the "unfound" dead, and questions the capacity of history, literature, or science to preserve their remains.[124] Throughout the fluctuating postbellum editions of *Leaves of Grass*, Whitman constructs a literary tomb that expands and contracts in an effort to house these diasporic ghosts.[125]

While Whitman sought to preserve the war's "human fragments," he also felt dissonant desires for their exhibition and concealment. *Memoranda* oscillates between protecting convalescent privacy and allowing readers to glimpse the hospital's "specimen interiors": "a hundred unnamed lights and shades of camp . . . will never be written—perhaps must not and should not be."[126] In a draft letter to Emerson dated January 17, 1863, Whitman described his inability to capture the hospital milieu in "current" language. A new lexicon was needed: "The first shudder has long passed over, and I must say I find deep things, unreckon'd by current print or speech."[127] Years later, he described to Traubel the guarded intimacy that he shared with soldiers as "the revelation of an exquisite courtesy—man to man—rubbing up there together: I could say in the highest sense, *propriety—propriety,* as in the doing of necessary unnameable things, always done with exquisite delicacy."[128]

Yet, the sexual history of specimens did not remain entirely "unwritten."[129] Whitman alludes to the existence of an actual interior where soldiers could retreat: the ward-master's room.[130] The poet frequently spent entire nights at the hospital, napping in a room adjacent to the ward-master's office.[131] Each ward contained a similar chamber, which afforded occupants a degree of privacy, provided the master was agreeable.[132] Hospitality was not guaranteed, and discretion was necessary. A letter from William Stewart to Whitman suggests that the poet had been chastised by a ward master: "I hope that don't keep you away, what the ward master told you that night."[133] Alluding to the erotic possibilities offered within the master's room, another soldier, Alonzo Bush, wrote to Whitman on December 22, 1863:

> I am glad to know you are once more in the hotbed city of Washington so that you can go often and see that Friend of ours at Armory Square, L.K.B. The fellow that went down on your BK, both so often with me. I wished that I could see him this evening and go in the Ward Master's Room and have some fun, for he is a gay boy.[134]

The initials L.K.B. almost certainly refer to Lewis ("Lewy") Kirk Brown, a twenty-year-old Maryland soldier who was hospitalized at Armory Square. Whitman described Brown as "a most affectionate fellow . . . very fond of having me come and sit by him."[135] Bush went on to express sympathy regarding Brown's impending amputation: "I am very sorry to here [sic] that after laying So long that he is about to loose his leg."[136] The enigmatic "BK" is open to interpretation, though both Roper and Katz demonstrate that "buck" was common slang for penis in nineteenth-century American vernacular.[137] The reference to Brown as "gay," while sexually suggestive, did not have an exclusively queer context; rather, the nineteenth-century usage indicated an individual willing to perform oral sex, often a female prostitute.[138] Bush was probably referring to the master's room on Ward K, which was supervised by Charles Cate, with whom Whitman and many of his favorite soldiers were on friendly terms.[139]

Like so many of Whitman's letters, Alonzo Bush straddles the border between privacy and lucidity, inhabiting Sedgwick's "glass closet," in which meaning is simultaneously concealed from potentially hostile readers, and revealed to an initiated audience.[140] Veiled descriptions of this erotic convalescent world may be rare because the participants' efforts at concealment were successful.[141] Alternatively, such acts may be largely undocumented because they were considered, by many, to be unremarkable.[142] Intense friendships between men, which often included physical affection, intimate touch, and bed-sharing, were common in antebellum America.[143] As Katz notes, Civil War camp sleeping arrangements made it easy for men-seeking men to find sexual partners. Whitman's notebooks and correspondence frame this "specimen interior" as both a "glass closet" and a cabinet of curiosities: an infinitely adaptable space in which to preserve the "unwritten history of the war."[144]

In postbellum editions of *Leaves of Grass*, Whitman memorializes countless "unfound" casualties, while guarding a traumatic history that remains "buried in the grave."[145] It is an inherently unstable text (most evident in Whitman's ongoing revision process), at times enacting what Sedgwick has termed "the speech act[s] of silence—not a particular silence, but a silence that accrues particularity by fits and starts, in relation to the discourse that surrounds and differentially constitutes it."[146] Whitman constructs a book that gestures toward the unspeakable, while remaining reverentially silent on that which "must not" be written.[147] Yet this silence is not without anxiety, as foreshadowed in the first edition: "What living and buried speech is always vibrating here, what howls restrain'd by decorum."[148] This restrained vibration suggests that the secret seeks its own revelation. Whitman's "buried speech" desires exhumation, in keeping with the phenomenon that Abraham and Torok have called "cryptonymy":

> Cryptonymy (coined from the Greek prefix *crypto* for "hiding" and an analogy with rhetorical terms such as met*onymy*) is a verbal procedure leading to the creation of a text . . . whose sole purpose is to hide words that are hypothesized as having to remain beyond reach.[149]

The cryptophoric subject is speechless in the face of trauma: articulation would destroy the illusion of integration with the lost other. Yet the process of encryption is neither final nor fail-safe.[150] The psychic crypt conceals traumatically provocative words, yet this "buried speech" does not remain mute, or still. The silences surrounding crypts betray their existence. Alive, they vibrate within the grave.

Whitman refashioned *Leaves of Grass* as a poetic tomb for the interment of histories that the author believed must remain "untold" but not unrepresented. As Nicholas Rand writes: "Words are manipulated by cryptonymy as dried flowers in a herbarium. . . . Cryptonyms create a collection of words, a verbarium, with no apparent aim to carry any form of knowledge or conviction."[151] Rand's reading of the cryptonymic collection as botanical herbarium draws obvious parallels with Whitman's specimens. However, Whitman *does* seek to convey knowledge, while simultaneously guarding collective losses that resist the capacity of speech. According to Abraham and Torok, cryptonyms are catalyzed by a traumatic failure of language: "*the words themselves, expressing desire, are deemed to be generators of a situation that must be avoided and voided retroactively.* . . . For this to occur, a catastrophic situation has to have been created precisely by words."[152] Whitman's "catastrophic situation" is the incapacity of language not only to capture the atrocities of war, but also the "unworldliness" of the soldiers he attended and the desire that arose between them.[153] For Whitman, certain words related to the war and its casualties must remain "buried in darkness," yet the author retains access to their histories, which he selectively translates, and their phantoms, whom he spectacularly revives. Whitman does not "retroactively" "void" the traumas of war.[154] Rather, the poet reframes traumatic memory not as a series of psychic intrusions upon the vault of repression, but as a mechanism for visitations with his beloved ghosts. The "specimen" becomes his cryptonym, the all-encompassing "magic word" that absorbs soldiers' lost bodies and the poet's unsevered attachments to them.[155] Such talismanic signifiers become, in Abraham and Torok's words, the subject's "inseparable companions, they create and recreate the poem of the tomb deep within. . . . From beyond the tomb, pleasure . . . nevertheless."[156] An extension of the libidinal surge that accompanies mourning, "magic words" are spiritualist ciphers of conjuration, offering the possibility of ongoing erotic contact with the dead.

It is not only the unutterability of certain events (whether traumatic, euphoric, or both) but also the "convergence of social prohibition and melancholia" that, according to Judith Butler, results in the burial of speech:

> We might ask what remains unspeakable here, not in order to produce speech that will fill the gap but to ask about the convergence of social prohibition and melancholia, how the condemnations under which one lives turn into repudiations that one performs, and how the grievances that emerge against the public law also constitute conflicted errors to overcome the muted rage of one's own repudiations. In confronting the unspeakable in Antigone, are we confronting a socially instituted foreclosure of the intelligible, a socially instituted melancholia in which the unintelligible life emerges in language as a living body might be interred into a tomb?[157]

In the aftermath of war, Whitman was driven not only to produce a form of speech that could "fill the gap" surrounding loss, but to do so within the constraints of societal prohibition about what kinds of attachment are named (or must remain unnamed). The poet's collected specimens speak to the melancholic arousal of an insatiable linguistic wound, which remains empty no matter how many words it contains.

In contrast to Whitman's invocation of "specimens" as "active, breathing forms," Brinton constructs a cabinet of curiosities in which the lost object is permanently quarantined.[158] The Army Medical Museum operates as a site of *false* incorporation, where the "foreign body" is not lodged within the psyche, but encased forever in its foreignness, entirely other and inaccessible. Incorporation invokes the psychic wound that Abraham and Torok call the crypt, where the lost other is *kept alive* within the ego.[159] The museum inverts incorporative mourning processes, creating an actual crypt where the lost object is *kept dead*, and externalized. It can never be reclaimed or reintegrated, not even by its "original possessor." As a model of false incorporation, the museum encloses its specimens within glass cabinets whose locks replicate lost human hands, visible yet hovering just out of reach.

The drive to preserve specimens as an inversion of psychic incorporation is epitomized in the work of Brinton's contemporary, the anatomist Joseph Leidy. A surgeon at Philadelphia's Satterlee Hospital during the war, Leidy contributed an unusually high number of specimens to the Army Medical Museum.[160] The notes accompanying specimens 93 to 98, all from the same soldier, reflect the poetic imagery of Leidy's autopsy rhetoric. Corporal G. S. of the 9th Wisconsin Regiment died on Christmas Eve, 1862.

> Autopsy: Age, about 30 years; body rather emaciated; abdomen presented a number of faint spots of purpura; lobular pneumonia in lower lobes of both lungs, the inflamed portions numerous, from the size of a marble to that of a walnut . . . Peyer's glands darkened with inflammation; solitary glands looked like yellow mustard seeds sprinkled on a red ground; large intestine streaked and spotted with ash-color and dark red on a more uniform red ground.[161]

From this description, a nonmedical audience can deduce that G. S. suffered from severe respiratory dysfunction, that his immune system was compromised by poor nutrition, and that he endured severe intestinal maladies. Simply put, the postmortem tells us that he died of complications arising from starvation and exposure. Echoing Whitman's description of gaunt Southern soldiers, many of Leidy's autopsy notes contain the same opening line: "body rather emaciated."[162]

Leidy's florid descriptions of the body's interior occupy the intersection of anatomy and botany: glands appear as mustard seeds scattered on red earth, intestines are streaked with ash. While Leidy was one of the era's most eminent anatomists, he was also a leading American naturalist. In 1853 he was appointed professor of human anatomy at the University of Pennsylvania Medical School, a post he held for the remainder of his life. Brinton's assistant, Fred Schaferdt, trained under his supervision.[163] Leidy's *Elementary Treatise on Human Anatomy* (1861) remained the authoritative text on the subject for over half a century, featuring over 300 illustrations personally drawn from observations made in the dissection laboratory at the University of Pennsylvania.

Whitman and Leidy were personally acquainted and shared several friends within the College of Physicians of Philadelphia, including Silas Weir Mitchell.[164] Like Whitman, Leidy often wrote drafts on scraps of old paper. In a pithy twist of intimate humor that would no doubt have delighted Whitman, Leidy's notes and sketches for an anatomical lecture on the human penis were written on the back of an invitation to a dinner lecture given by Whitman at the Philadelphia Contemporary Club, signed in the poet's own hand.[165]

During the latter half of the nineteenth century, the surgeon's art of dissection fused with the compulsion to preserve human remains in the practice of surgeons and bibliophiles who collected books bound in human skin, a technique known as anthropodermic bibliopegy. The creators of such volumes were usually medical men, many of whom flayed and tanned the skins themselves. Leidy bound his personal copy of *An Elementary Treatise on Human Anatomy* in the skin of a Civil War soldier. Outwardly, hide-bound books keep their secrets well; human leather does not betray its origins to the naked eye. The visual link between skin bindings and the bodies they once covered was erased in the transformation from tissue to text. In order to identify their work, handwritten inscriptions signed by the doctor or collector document the books' human exterior.[166] Leidy inscribed the front flyleaf of his volume: "The leather with which this book is bound is human skin, from a soldier who died in the great southern rebellion." The binding is soft beige leather, bearing no trace of the nameless soldier from whom it was sourced.

Leidy's copy of his masterwork captures the anatomical impulse of inverted incorporation: the surgeon appropriates his patient's skin to enclose his own theories on the human body. The book assumes the body's protective sheath, binding this anatomical treatise to one specific cadaver. Leidy's book is unique in the sense that, according to museum records, the doctor conceived of it as an homage to the slain soldier. As Carolyn Marvin's history of human-leather books establishes, like the cadavers used for dissection, skin bindings were almost always sourced from marginal bodies:

> The human hide-bound book was situated in the scholar's library and the museum, entitled private places. . . . What made it morally defensible to use the poor in this fashion . . . was the social construction of the indigent and outcast as surplus people . . . useful at least for medical science, or, in the case of the bound book, for indulging the esoteric intellectual interests of medical practitioners. . . . Like medical body snatching, it made of the poor a resource for the production of physicians.[167]

In spite of its commemorative intentions, Leidy's handiwork remains marked by hierarchies of privilege. The soldier whose skin he repurposed apparently did not have the means to arrange for his body to be embalmed and repatriated home. Leidy worked closely with Brinton at the College of Physicians of Philadelphia, where both men advocated for the Pennsylvania Anatomy Act. Drafted by the College of Physicians in 1867, the act's full title unites medical advancement with the prevention of body snatching: An Act for the Promotion of Medical Science, and to Prevent the Traffic in Human Bodies.[168] Despite the legislation's passage, medical colleges continually struggled to obtain cadavers. In a letter to the Philadelphia Board of Charities, Brinton complained:

> It is true that by the Provisions of the Anatomy Act unclaimed bodies of persons dying in a mortuary are turned over to the medical colleges . . . [Yet] the interests of medical teaching in the City of Philadelphia are seriously threatened by the scarcity of suitable material for dissection, and that scarcity is yearly increasing.[169]

In the postbellum years, anatomical appropriation not only continued to thrive, it was privately celebrated by medical men. In an 1891 memorial address to the University of Pennsylvania medical department, Dr. William Hunt recalled a bizarre tale of medical theft undertaken by Leidy in order to acquire two petrified bodies. This story demonstrates the personal risks doctors were prepared to take in order to secure rare specimens:

> Some years ago [Leidy] came to my house in quite an enthusiastic mood, and said, "Dr. Hunt, do you know that they are moving the bodies from a very old burying-ground down town to make way for improvements?" "Yes," I said. "Well," he went on, "two bodies turned into adipocere are

> there (this is an ammoniacal soap, and the bodies are commonly called *petrefied bodies*). They have been buried for nearly a hundred years, an old man and an old woman; nobody claims them, and they would be rare and instructive additions to our collections. Now, I think I can get them, and you will take one for the Mütter Museum, I will take the other for the Wistar and Horner Museum." All right, I said, I shall be delighted. So down Leidy went, full of the idea of securing the prizes. When he spoke to the superintendent or caretaker of the ground, that gentleman put on airs, talked of violating graves, etc.; so the discomfited doctor was going away quite crestfallen. Just then the caretaker touched him significantly on the elbow and said, "I tell you what I do; I give bodies up to the order of relatives!" The doctor immediately took the hint. He went home, hired a furniture wagon, and armed the driver with an order reading, "Please deliver to the bearer the bodies of my grandfather and grandmother." This brought the coveted prizes, and one is now in each of the museums, and the virtuous caretaker was amply compensated.

Alongside Leidy's anatomical treatise, the body of the petrified woman that he removed from the cemetery remains on display in the Mütter Museum to this day.[170] Speaking to an audience of medical colleagues, Hunt readily admitted that deception and bribery were integral to the procurement of these "coveted prizes." His eulogy concludes with a verse written to explain to a curious friend the exhumation compulsion that compelled surgeons and their students:

> I had occasion to write some lines to a lady, not long since, one who has much interest in a certain mortuary, . . . about this peculiar tendency in men medical. I admit the fact, you will notice, and discuss the law—that is, the metaphysics of it, for I believe jurists have frequently declared there is no property in a dead body.
>
> Now I'll tell you a secret,
> That makes me grieve!
> The fellows who post
> Will always thieve!
>
> Though the *corpus delicti* is not quite clear,
> For 'tis plain a dead body can't say, "I am here,"
> Or "It is I," "*homo sum*," "me too," or "Ego,"

> For all are agreed he's above, or below.
> So to the posters don't worry as to *tuum* and *meum*,
> But take the specimens for the museum.[171]

Hunt surpasses the seventeenth-century anatomist William Shippen's argument of bioethical privilege, laid out in chapter 1, that the bodies of "paupers and the friendless dead" were a fit source of raw material, while those buried in churches and private burial grounds should remain unmolested.[172] For Hunt, the privacy of interment matters not, for "there is no property in a dead body." The logic that ownership does not apply to the corpse echoes Brinton's dismissive response to the soldier who wished to reclaim his severed limb. A "dead body" cannot claim, "*homo sum*," which translates as "I am a man." The later, omitted half of this quote, which was likely assumed knowledge among Hunt's audience, is eerily revealing: *Homo sum: humani nihil a me alienum puto*; "I am a man: I deem nothing pertaining to man foreign to me."

These words, written by the playwright P. Terentius Afer, better known as Terence, echoed across the Roman Empire in the mid-second century BC. According to historian Robert A. Bauman, "the influence of Terence's felicitous phrase on Roman thinking about human rights can hardly be overestimated." As an African and a freed slave, Terence embraced the philosophy of universalism, "the essential unity of the human race."[173] Two hundred years later, the philosopher Seneca closed his exposition on humanism with Terence's famous maxim:

> There is one short rule that should regulate human relationships. All that you see, both divine and human, is one. We are the parts of one great body. Nature created us from the same source and to the same end. . . . Let that well-known line be in our hearts and on our lips: Homo sum: humani nihil a me alienum puto.[174]

Seneca's words resonate in Whitman's corporeal and spiritual synecdoche: all parts of the body are divine, all humans are part of "one great body." Thus, no part of that body (alive or dead, attached or detached) can be seen as foreign to any other human being: "O I say these are not the parts and poems of the Body only, but of the Soul."[175] Beyond this philosophical convergence, the political landscape in which Terence wrote bore other striking similarities to nineteenth-century America. Driven by the need to delineate their relations with non-Romans across the empire, thinkers such as Cicero and Seneca articulated the theory of *humanitas*. Influenced by the Greek tenets of *philanthropia*, its primary concern was the ethics instilled in citizens through education. As Bauman observes in his history of human rights in ancient Rome, *humanitas* functioned as an "incentive to avoid brutal behavior towards other members of the human race, either as an individual or in groups."[176] Given this contextualization, Hunt's truncated invocation of Terence's words to defend the "peculiar tendency" of medical body snatching is all the more ironic, linking the theft of human cadavers with the "peculiar institution" of slavery.

Hunt's satirical defense of body snatching hinges on the mute corpse, who cannot speak to claim her lost humanity. This posthumous silence connects Hunt's stolen specimens with Whitman's "unfound" soldiers and the histories that remained buried in (or resurrected from) their graves.[177] In contrast to Leidy's anthropodermic treatise as an object of inverted incorporation, where human tissue binds corpse to text, *Leaves of Grass* enacts symbolic burials for countless anonymous specimens. While he did not actually decorate his "leaves" with their bodies, Whitman's "reminiscent memorial" remains an external form of incorporation through which the book, rather than the mind, entombs an ambiguous loss too vast for one man alone to contain.[178]

The Noblest Specimen

Skeletal remains from at least four of Whitman's soldiers became artifacts in the Army Medical Museum.[179] All four were submitted by Dr. D. Willard Bliss, chief surgeon at Armory Square Hospital, whom Whitman described as "one of the best surgeons in the army."[180] Bliss reciprocally praised Whitman for his devotion to the wounded: "No one person who assisted in the hospitals during the war accomplished so much good to the soldier and for the Government as Mr. Whitman."[181] Did Whitman know that remains of these soldiers were displayed in the museum? Did he visit them there? While we may never definitively answer these questions, we do know that divergent forms of preservation (textual and medical) intersected upon the bodies of these four men, who were nursed by Whitman in life and curated by Brinton after death.

Oscar Cunningham was shot in the right leg at the Battle of Chancellorsville on May 2, 1863. He was admitted to Armory Square on June 10, where the bullet was extracted on June 15.[182] Shortly after Cunningham's arrival, Whitman observed, "I thought he ought to have been taken to a sculptor to model for an emblematic figure of the west, he was such a handsome young giant over 6 feet high, with a great head of brown yellow hair. . . . He has suffered very much since—the doctors have been trying to save his leg but it will probably have to be taken off yet."[183] Whitman's prediction proved accurate. Abscesses surrounded the incision, necessitating amputation at the thigh. Bliss performed the surgery on May 2, 1864, the anniversary of Oscar's initial wounding.[184] Whitman mourned the deterioration of his "youthful physical manliness" over the course of that year:

> I have just left Oscar Cunningham . . . he is in a dying condition . . . it would draw a tear from the hardest heart to look at him—he is all wasted away to a skeleton & looks like some one fifty years old—you remember I told you a year ago, when he was first brought in, I thought him the noblest specimen of a young western man I had seen . . . O what a change.[185]

Cunningham was still alive on May 5, when Bliss submitted his right femur to the museum, where it was cataloged as Surgical Specimen 2254. Surviving records do not indicate whether the soldier was aware that his limb was donated. Although Bliss's letter suggests that he held out hope for the patient's recovery, Cunningham died on June 5, 1864.[186] The case history contains a haunting reflection on the calcified traces of Cunningham's deterioration: "the new bone formation firmly retains the fragments, and is sufficiently rounded to indicate the lapse of considerable time."[187]

Whitman memorialized Oscar Wilbur as "A New York Soldier" in *Memoranda.* Wilbur sustained a "compound fracture of the femur" at the Battle of Chancellorsville on May 3, 1863.[188] (See figure 4.) He lay unattended on the battlefield for ten days before he was finally evacuated to Aquia Creek Hospital, where he remained for a further forty-two days before being transferred to Armory Square. According to Bliss's reports, he suffered from constant nausea and died of "exhaustion, July 31, 1863."[189] Whitman remembered him as a stoic, spiritual young man who reciprocated the poet's affections:

> I have spent a long time with Oscar F. Wilber, Company G, One Hundred and Fifty-fourth New York, low with chronic diarrhea, and a bad wound also. . . . He talk'd of death, and said he did not fear it. . . . He behaved very manly and affectionate. The kiss I gave him as I was leaving he return'd fourfold.[190]

The concept of the Good Death was vital to mid-nineteenth-century mourning practices. A peaceful death was as significant as an honorable life, perhaps

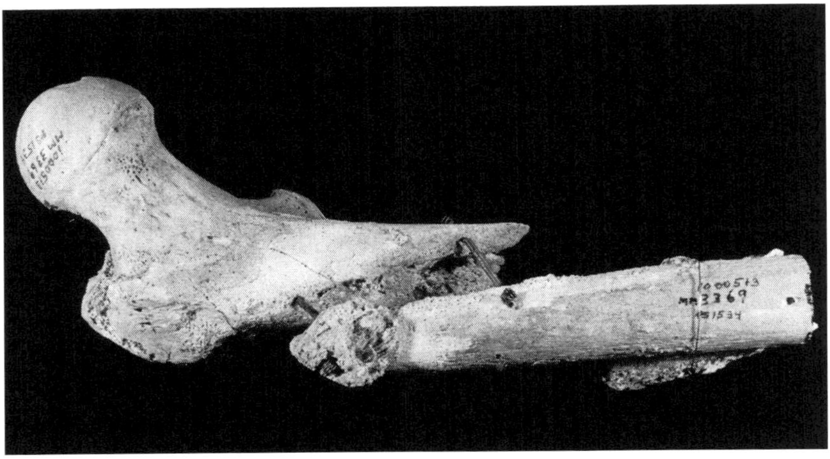

4. The upper right thighbone of Oscar Wilbur. National Museum of Health and Medicine.

even more so, as it was believed to foreshadow the status of the soul in the afterlife. Deathbed vigils required familial witnesses to observe the transition of the dying spirit. The violent, often isolated deaths of soldiers could not have been further removed from these domestic mourning rituals.[191] Whitman ensured that his specimen-soldiers did not "die among strangers without having one at hand who loved . . . [him] dearly."[192] The poet's deathbed presence allowed him to act as a surrogate mourner, to receive the soldier's "dying kiss." Oscar Wilbur died "a few days after" this encounter with Whitman.[193] His right femur was posthumously amputated for the purpose of donation to the museum, where it became Surgical Specimen 1534.

Whitman's most medically famous specimen was John Mahay, who was shot in the groin at the Second Battle of Bull Run on August 29, 1862. He was treated at Armory Square, where Whitman often visited him. Upon hearing of Mahay's unusual wound, Brinton personally interviewed the soldier. He kept a detailed account of Mahay's symptoms, including the passage of bone fragments through his urethra. Mahay died on October 24, 1862. During the autopsy, several urinary stones were removed from his bladder and cataloged as Surgical Specimen 2567.[194] Whitman memorialized the tragic narrative of Mahay's life and death in *Memoranda:*

> Well, poor John Mahay is dead. . . . His was a painful and lingering case. . . . The bladder had been perforated by a bullet going entirely

> through him. . . . Poor Mahay, a mere boy in age, but old in misfortune. He never knew the love of parents, was placed in his infancy in one of the New York charitable institutions, and subsequently bound out to a tyrannical master . . . the scars of whose cowhide and club remained yet on his back. . . . He found friends in his hospital life, and, indeed, was a universal favorite. He had quite a funeral ceremony.[195]

After enduring years of abuse, Mahay found a familial community in the ward. Posthumously, he attained medical notoriety. Unlike Whitman's other specimens, who were relegated to mere statistics, Mahay's unique wound earned a detailed description in the *Medical and Surgical History*, including graphic illustrations. The entry includes statements from both Bliss and Brinton, documenting the extent of the patient's suffering. Brinton observed, "[he] complains of pain at the anterior wound when he draws a long breath, and of constant pain in the glans penis." Bliss stated, simply, "he has never been perfectly free from pain."[196]

On March 25, 1865, Frank H. Irwin sustained a gunshot wound in the left knee at the Battle of Fort Fisher. Three days later he was transferred to Armory Square, where Bliss amputated his leg on April 14. Assistant Surgeon M. J. Munger submitted his femur to the museum two days later.[197] The bone was cataloged as Surgical Specimen 4077.[198] This medical narrative offers no insight into the soldier's experience of surgery or its aftermath. Like Cunningham, Irwin was still alive when his specimen was submitted. He died on May 2, 1865, following a severe pyemia infection.[199] (See figure 5.)

In stark contrast to the neutrality of his medical history, Whitman recorded the tragic details of Irwin's demise under the heading "Death of a Pennsylvania Soldier." This entry reprints his condolence letter to Irwin's mother, words of comfort from a "casual friend that sat by his death bed":

> He seemed quite willing to die . . . I do not know his past life, but I feel as if it must have been good. At any rate what I saw of him here, under the circumstances, with a painful wound, and among strangers, I can say that he behaved so brave, so composed, and so sweet and affectionate, it could not have been surpass'd. . . . I thought perhaps a few words, though from a stranger, about your son, from one who was with him at the last, might be worth while—for I loved the young man, though I but saw him immediately to loose him.[200]

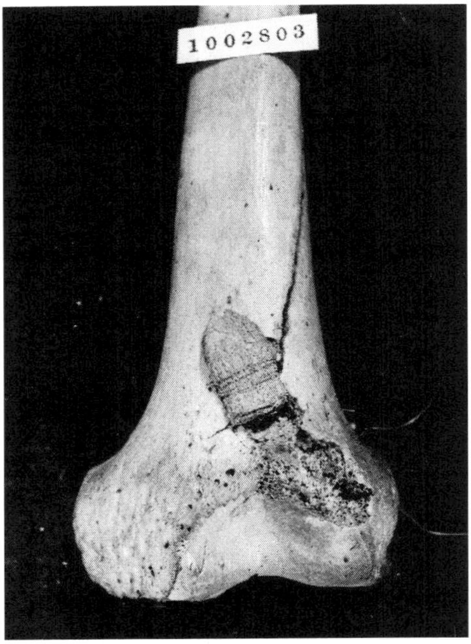

5. The lower left thighbone of Frank Irwin. National Museum of Health and Medicine.

Whitman frequently composed letters on behalf of soldiers and wrote expressions of sympathy to bereaved families.[201] He recalled the peculiar intimacies entailed in proxy letter writing: "I encourage the men to write, and myself, when call'd upon, write all sorts of letters for them (including love letters, very tender ones)."[202] Privileged antebellum deaths unfolded within the family home; hospitals were the last resort of the indigent or victims of epidemic disease. The *hors mori*, the "hour of death," was of paramount importance. The Civil War shattered these conventions, as soldiers died violently far from home. The testimony of a South Carolina woman in 1863 attests to the cultural perception of wartime death as diaspora, stating that it was "much more painful" to lose "a loved one [who] is a stranger in a strange land."[203] Letters of condolence allowed families to vicariously experience, through the narration of a proxy mourner, the moment of death. Soldiers sought out surrogates who could stand in for their absent families; nurses allowed delusional patients to believe they were their mothers, sisters, or wives. In the absence of an effective military system for reporting casualties, it was customary for the dead soldier's comrades or hospital attendants to write to his next of kin, not only to express sympathy, but also to provide evidence that their loved one's passing was in keeping with the conventions of dying well.[204] Condolence letters from the war period are remarkably similar, as their authors understood the facets of the Good Death, and sought to assure the bereaved that the soldier's final hours complied. Whitman's letter assures Irwin's family that their son's death was overseen by a loving witness, albeit a stranger. He attests to the soldier's "willingness to die," and praises his composure while suffering "a painful wound . . . among strangers."

Whitman articulates the stranger's capacity to mourn absolutely and immediately, in the absence of all prior knowledge ("I do not know his past life"). A letter to the family of Erastus Haskell, dated July 27, 1864, reflects the depth of the poet's deathbed attachments:

> From the time he came to Armory Square Hospital till he died, there was hardly a day but I was with him a portion of the time. . . . Many nights I sat in the hospital by his bedside till far in the night . . . I shall never forget those nights, it was a curious and solemn scene, the sick & wounded lying around in their cots, just visible in the darkness, & this dear young man close at hand lying on what proved to be his death bed—I do not know his past life, but what I do know & what I saw of him, he was a noble boy—I felt he was one I should get very much attached to. So farewell, dear boy—it was my opportunity to be with you in your last rapid days of death—no chance, as I have said, to do anything particular, for nothing could be done—only you did not lay here & die among strangers without having one at hand who loved you dearly, & to whom you gave your dying kiss.[205]

This letter narrates the soldier's decline alongside the poet's intensifying "attach[ment]," from the moment Haskell arrived in the hospital until the day he died. Whitman opens with a description of the "curious and solemn scene," allowing the reader to vicariously wander the wards, finally coming to rest at the deathbed of Erastus Haskell. Once again, Whitman attests to the soldier's nobility, though he knows nothing of his "past life." The deathbed alone provides evidence of the young man's character, and positions him as "one" of the "specimens" to whom Whitman was "very much attached." In closing, the poet departs from convention and addresses the deceased soldier directly. Whitman describes Haskell's "rapid days of death" as "my opportunity," framing the soldier's dying kiss as a final "gift."

Whitman's deathbed vigils demonstrate the capacity of strangers to form a binding affiliation with the dying, to mourn their passing even in the absence of personal history. The neglected symbol of the specimen in Whitman's writing reveals volumes about the intimacy of mourning strangers in nineteenth-century America. Moving beyond a necrophilic attachment to the corpse, the specimen recalls the allure of the phantom limb: an entity felt most acutely in its vacancy. As a signifier of embodied mourning, it traces connections that remain impossible to sever.

Whitman and Brinton offer convergent histories of the war's strange cases and dual bodies: military and corporeal, phantom and physical. Brinton's specimens are remnants of a fractured army, reconfigured within the museum to demonstrate the enduring coherence of the Union. Whitman insists upon the continual erotic relevance of absent bodies and their abandoned parts. His specimens are textually entombed, a synecdochic representation of bodies that eluded burial. In contrast, Brinton's specimens are perpetually unburied, preserved behind glass cases. Yet, both poetic and surgical collections insist that the Union endures, in spite of its wounds. The sutured democratic nation absorbs its "rejected members"—in Whitman's words, it "contain[s] multitudes."[206] When Whitman writes of a specimen, he articulates a physical, psychological, and often sexual connection to another human being. The Whitmanian body is more than the sum total of its parts, and the dismembered part is therefore more than the subtraction of its original whole. The body *is* the soul, and the amputated limb retains a trace of the spirit that inhabited the living form, intact.

The museum's shelves and the book's pages display carefully preserved afterlives of science and spirit. Brinton insisted that the museum was not merely a cabinet for war curiosities, but a national medical legacy: "the foundation of a great National Surgical and Medical Museum was not for the collection of curiosities, but for the accumulation of objects and data of lasting scientific interest."[207] Yet, as the curator's account of one soldier's attempted repossession demonstrates, these specimens are more than clinical "objects." The public display of human remains elicited traumatic reverberations in both soldiers and civilians. While the museum's twenty-first-century incarnation attests to the success of Brinton's project, the contemporary accounts analyzed in this chapter show that nineteenth-century audiences experienced dissonant affective responses to the collection.

Whitman's fascination with amputation paralleled his own desire to resurrect poetic incarnations of his lost specimens. As the following chapter reveals, nostalgic longing and phantom limbs both represent the irreconcilability of the past embodied within the subject. For Whitman, mourning is an ongoing attachment that transcends the physical presence of the body. The poet demonstrates an open-ended, queer hospitality to soldiers' "sensory ghosts" and their unhealing wounds.[208] The lingering influence of these absent limbs upon survivors and scientists frames the aftermath of amputation as a form of haunting that beset not only amputees, but also their caregivers, doctors, collectors, and voyeurs. These lost members possess their own afterlives, independent of the bodies they left behind.

3 Phantoms of Countless Lost
The Nostalgia of Absent Limbs

In January 1864 Whitman witnessed the amputation of Lewis Brown's left leg. The young private was shot at the Second Battle of Bull Run on August 21, 1862. The wound remained unhealed after sixteen months at Armory Square Hospital, where Brown and Whitman became intimate friends.[1] Chief Surgeon Willard Bliss performed the operation—the same doctor who submitted specimens extracted from Oscar Cunningham, Oscar Wilbur, John Mahay, and Frank Irwin to the Army Medical Museum.[2] Whitman watched from the doorway as Brown's leg was severed just below the knee, recording the scene in his notebook:

> The surgeon in charge amputated but did not finish the operation, being called away as [the attendant] was stitching it up. Lewy came out of the influence of the ether. It bled and they thought an artery had opened. They were ready [to] cut the stitches again & make a search but after some time concluded it was only surface bleeding. They stitched it up again and Lewy felt every one of those stitches though yet partially under the influence of ether. They did not think it safe to give him any more as he had already taken it excessively. I could hear his cries sometimes quite loud & half-coherent talk & caught glimpses of him through the open door. At length they finished & they brought the boy in on his cot . . . I sat down by him. . . . His face was very pale, his eyes dull. . . . He remained very sick, opprest for breath, with deathly feeling.[3]

Similar in circumstance to many of his comrades, Brown endured surgery with insufficient anesthesia and was partially conscious during the severance of his leg.[4] In the following days, Whitman was never far from his bedside, sleeping on the adjoining cot, and documenting the amputation's neurological consequences: "As usual in such cases [Lewy] could feel the lost foot & leg very plainly. The toes would get twisted, & not possible to disentangle them."[5] Whitman's characterization of Lewy's ghostly contractions as "usual in such cases" is a remarkable testament to his own powers of observation and his extensive experience with war casualties; this is no mere reference to an established disorder. At the time of his writing, "phantom limb" did not exist as a medical phenomenon. In 1551 the famed French surgeon Ambroise Paré observed a curious affliction in his patients, a description that is considered by medical historians to be the first account of phantom limb: "Verily it is a thing wondrous strange and prodigious, and which will scarce be credited, unless by such as have seene with their eyes, and heard with their ears the Patient who have many moneths after the cutting away of the legge, grievously complained that they yet felt exceeding great paine of that Leg so cut off."[6] After Paré's account, the phantom vanished from medical literature for 320 years, until the unrelenting spate of amputations during the Civil War inspired the Philadelphia neurologist Silas Weir Mitchell to resurrect this "unseen ghost."[7] Like other invisible, but no less real, traumas experienced by Civil War soldiers, phantom pain existed outside corporeal borders, in severed nerves that remained enmeshed—impossible to untangle.

In the years between Fort Sumter and Appomattox, more than 750,000 soldiers died, while another 500,000 (at least) were injured—a human cost equal to all other U.S. conflicts combined.[8] The primary catalyst for these alarming statistics was the minié ball, a conical bullet whose velocity drastically enhanced the severity of gunshot wounds.[9] Hands and fingers were the most commonly severed extremity; knee, hip, shoulder, and elbow amputations were also frequent.[10] Several years before comprehensive casualty statistics were released, Whitman observed this same configuration of injuries: "A large majority of the wounds are in the arms and legs. But there is every kind of wound, in every part of the body."[11] Speed was essential to survival; unlike Brown's prolonged surgery, most amputations were over within two minutes, including closing the arteries and suturing skin over the stump.[12] Due to the absence of antiseptic precautions, septicemia and gangrene led to high mortality rates. Whitman described the casual atmosphere in the operating theater: "the principal here, Dr. Bliss, is a very fine operating surgeon—sometimes he performs several amputations or other operations of importance in a day—amputations, blood, death are nothing to him—you will see a group absorbed in playing cards up at the other end of the room."[13] Whitman's account of Bliss's surgery admits the necessity of medical detachment, while illuminating the intimate camaraderie between soldiers. The wards were scenes of countless traumas, but they were also social spaces of masculine domesticity where soldiers resided in close proximity, sometimes for months or even years.

Whitman devoted his hospital years to assuaging the "pangs of aggravated wounds" that afflicted soldiers in mind and body.[14] *Drum-Taps* and its subsequent incarnations within *Leaves of Grass* remain attuned to the war's "sensory ghosts," both physical and psychological.[15] In "The Wound-Dresser" Whitman describes the visual trauma of dismemberment:

> From the stump of the arm, the amputated hand,
> I undo the clotted lint, remove the slough, wash off the
>   matter and blood,
> Back on his pillow the soldier bends, with curv'd neck,
>   and side falling head;
> His eyes are closed, his face is pale, he dares not look on
>   the bloody stump,
> And has not yet looked on it.[16]

While the "dresser" removes his bandages, the soldier turns away, unable to look at what remains of his arm. Paradoxically, the poet's lingering gaze allows the reader to glimpse a wound so visceral that it remains unseen by the amputee. From the Civil War until the present day, doctors have theorized that the phantom limb's invisibility is central to its bearer's suffering. In an era when more surviving amputees are returning from current U.S. wars than during all of the twentieth century's conflicts, neurologists are once again striving to ease the limbless perceptions of escalating military casualties.[17] Illustrating the prophetic empathy of Whitman's convalescent poetics, mirror images that replicate ethereal limbs have become effective therapeutic tools. The visual integration of sensory experience often results in the decrease or cessation of phantoms.[18]

The medical history of phantom pain is rife with contradiction and controversy. Resistance to the elusive phenomenon of disembodied perception caused sufferers to remain clinically marginalized well into the twentieth century.[19] As V. S. Ramachandran and William Hirstein observed: "Although patients with this syndrome have been studied extensively since the turn of the century, there has been a tendency among physicians to regard them as enigmatic, clinical curiosities."[20] Most postbellum doctors assumed that phantoms originated from remaining peripheral parts, residing in scar tissue and nerve endings.[21] For much of the twentieth century, nerve irritation theorists believed that nerve severance followed by partial regeneration was the only explanation for phantom limbs.[22] This theory was discredited as research revealed that phantoms are often strongest immediately after amputation, and most returned even after corrective surgery to amend extant nerves.[23] Twenty-first-century neurology has discarded the flawed model of the phantom limb as a spectral simulacrum of the intact extremity. As Cassandra S. Crawford articulates, "phantoms are today conceived as parts not accountable to gravity, symmetry, time, or the principles of human morphology, not answerable to the laws that had always governed the physiology of human bodies."[24] Contemporary science confirms what Civil War soldiers intuited: the phantom is its own entity, no longer beholden to its formerly physical incarnation. This state of embodied partiality has baffled medical researchers for centuries, in no small part because it defies temporality as well as physicality, recalling Hamlet's utterance in the wake of his father's ghost, "time is out of joint." This chapter offers the first critical analysis of Whitman's representations of phantoms alongside Civil War doctors' and amputees' narratives of "spirit members" that return, even after long absences, to haunt their hosts' bodies in dreams and waking life.[25]

Deathly Feeling

The presence of disfigured soldiers throughout Whitman's postbellum poetry and prose attests to the uncanny allure of the war's "rejected members" and their perpetually open wounds.[26] His hospital notebooks document the neural legacies of amputations, such as the case history of "Thomas H. B. Geiger, co. B 63rd Penn, wounded at Fredericksburg . . . young, bright, handsome Penn boy—tells me that for some time after his hand was off he could yet feel it—could feel the fingers open and shut."[27] In *Drum Taps* and *Memoranda during the War* Whitman narrates an enduring fidelity to maimed and partial bodies, perhaps most evident in his devotion to amputees and their lost parts. In her work on traumatic history, Cathy Caruth describes this type of attachment as an "encounter with another, through the very possibility and surprise of listening to another's wound."[28] Whitman mused on the "reciprocal love" that passed between men living and dying in the hospitals' "peculiar conditions":

> I met in the hospitals the manliest and tenderest characters. I loved them all. Describing the peculiar conditions in which love, compassion, reciprocal love, are drawn out. . . . I will tell you of one case: Virginian,

somewhat over 30 years old, had moved to Tennessee and could not read or write, had a full, somewhat bloodless face, large blue eyes, well apart, and perfectly formed features—the expression of some calm, beautiful animal.[29]

Whitman's specimen soldiers possess an otherworldly magnetism, marked by spectral "beaut[y]" and "animal purity."[30] Their animality is not inhuman, but rather an intensified form of embodiment, catalyzed by a suffering so acute that it borders on transcendent. Far from creatures of unbridled physicality, Whitman's specimens are inherently "spiritual."[31] This fusion of spectrality and animality was influenced by the work of the Viennese physician Franz Anton Mesmer, founder of the theory that bears his name, mesmerism. He argued that physical and spiritual phenomena were united by a magnetic current called the odic force.[32] Mesmer and other nineteenth-century medical practitioners believed that this force resided within the human body, and that certain people were especially receptive to the magnetism exuded by others.[33] Mesmer's technique of "animal magnetism" was promoted as a powerful vehicle for physical healing, where the practitioner directed his own energy through the body of the sufferer to induce a restorative trance.[34] Whitman believed that his "Presence" was more therapeutically effective than either traditional nursing or the tokens he bestowed upon wounded soldiers. To "prepare" for the "romance of surgery," Whitman cultivated his own body as a magnetic channel:[35]

> In my visits to the Hospitals I found it was in the simple matter of Personal Presence, and emanating ordinary cheer and magnetism, that I succeeded and help'd more than by medical nursing, or delicacies, or gifts of money, or anything else. During the war I possess'd the perfection of physical health. My habit, when practicable, was to prepare for starting out on one of those daily or nightly tours, of from a couple to four or five hours, by fortifying myself with previous rest, the bath, clean clothes, a good meal, and as cheerful an appearance as possible.[36]

Within this convalescent realm, the specimen soldier represents a figure of magnetic reciprocity, a conduit for "unworldliness."[37] Writing to a New York friend in March 1864, Whitman described the compelling tenderness that drew him to the wards: "My hospital ministrations are very fascinating with all of their sadness. The wounded & sick get incredibly near to one. Poor young men, they respond so affectionately to kindness & magnetism."[38] Initially captivated by the proximity of death, Whitman's continued presence is bound by the soldiers' receptivity to his ministrations. Like the ghostly presence of absent limbs, Whitman's specimens occupy the threshold between the living and the dead—"unconscious" and possessed by "the calmness of heaven."[39] Like Lewy Brown's lost toes, for Whitman the body and the soul are "impossible to disentangle"—each particle of the human animal embodies its "spiritual character."[40] What happens, then, to the spiritual traces attached to remnants severed from bodies still living? Whitman was not the only one to ponder this question. Civil War doctors were inundated with tales of men haunted by their lost parts.

The frequency of Civil War amputations led to an epidemic of phantom limbs, a diagnostic term conceived by Whitman's friend and physician, Silas Weir Mitchell, who was known as the "father of American neurology." He coined the phrase in an 1871 article that offered the first modern medical account of the phenomenon he described as a "sensory ghost."[41] Recalling Whitman's portrayal of the soldier unable to turn his gaze toward his partial arm, Mitchell described amputees "haunted by a constant or inconstant fractional phantom of so much of himself as has been lopped away—an unseen ghost of the lost part."[42] Mitchell's Civil War service led to a fascination with the consequences of gunshot wounds. As the war drew to a close, Mitchell returned to private practice. While working as a full-time physician, he remained a prolific medical researcher and writer, authoring more than 100 articles and books related to neurology, toxicology, pharmacology, and physiology.[43] He also wrote poetry and prose, much of which centered on the war.[44] Mitchell's 1871 book, *Wear and Tear, or Hints for the Overworked*, unveiled his infamous "Rest Cure" for nervous maladies, which prescribed seclusion, massage, electrotherapy, and a high-fat diet. Over the following decades, the "Rest Cure" was widely endorsed by the American medical establishment and embraced by Sigmund Freud and Jean-Martin Charcot, luminaries in the emerging field of psychology.[45] In the postbellum years, Mitchell became Whitman's friend and patron as well as his doctor.

Whitman's hospital work placed him in constant contact with contagious diseases and infected wounds. In the summer of 1863, his hand was nicked while assisting with an amputation.[46] The wound became infected; angry red lines traveled up to his shoulder, followed by chronic headaches and fever.[47] Yet there were debilitating symptoms even before the accident, telltale signs of trauma, exposure, and exhaustion: "a bad humming feeling and deafness, stupor-like at times."[48] While Whitman apparently recovered from this infected wound, the surgeons he encountered daily warned against "hospital fatigue" and the dangers of inhaling viral miasmas.[49] As Loving establishes, Whitman contracted tuberculosis during the war years, a condition that would compromise his famous vitality for the remainder of his life.[50] By the spring of 1864, doctors strongly advised the poet to avoid the hospitals, following spells of weakness and cranial pressure. In June of that year Whitman returned to Brooklyn, where he convalesced for six months. In a letter to William O'Connor, Whitman expressed a restrained faith in his resilience: "My physician thinks that time, with the change of locality & my own recuperative power, will make me well, but says my system is probably saturated with the virus of the hospitals &c which eludes ordinary treatment."[51] Whitman's conception of his illness as "saturation" illustrates his literary and corporeal incorporation of the war and its casualties into his book and his body. In a letter to his brother Jeff, the poet explained his "permanent" cathexis to "my hospitals":

> I cannot give up my Hospitals yet. I never before had my feelings so thoroughly and (so far) permanently absorbed, to the very roots, as by these huge swarms of dear, wounded, sick, dying boys—I get very much attached to some of them, and many of them have come to depend on seeing me, and having me sit by them a few minutes, as if for their lives.[52]

Confirming the depth of his attachment to the soldiers, in 1865 Whitman returned to Washington as soon as he recovered, and continued his hospital work until the final wards closed in 1866.[53] Like many of the men he attended, the poet would never entirely recover from the traumas of war, both physical and psychological. In 1873 Whitman suffered a stroke at the age of fifty-four.[54] He relocated from Washington, DC, to Camden, New Jersey, to live with his brother George. He eventually acquired the home on Mickle Street where he spent the remaining years of his life in declining physical health, all the while continuing to revise *Leaves of Grass*, and accruing a devoted cohort of disciples.

Whitman first consulted Silas Weir Mitchell on April 18, 1878, to discuss partial paralysis resulting from successive strokes.[55] In a letter to his sister-in-law Louisa, the poet wrote that a friend had driven him "over in the coupé to Philadelphia," where he "saw the great Dr. Mitchell, I was very well pleas'd with him—I am to go again."[56] On his second visit, Mitchell diagnosed the poet's paralysis as the result of a ruptured blood vessel in the brain, but pronounced Whitman's heart "normal and healthy."[57] Mitchell was the first physician to observe the psychosomatic features of Whitman's symptoms. After his second visit, Whitman recorded that Mitchell believed "the *bad spells*" to be "recurrences by habit (sort of automatic)."[58] The neurologist prescribed fresh air and sunbaths, a remedy that Whitman already wholeheartedly embraced.[59] Mitchell's fictionalized version of this encounter in *Dr. North and His Friends* (1910) sheds light on their rapport. During the consultation, Mitchell reportedly counseled Whitman to "live outdoors and take no physic." When Whitman inquired as to the fee the neurologist replied, "The debt was paid long ago; it is you who are still the creditor."[60] While the two writers did not always see eye to eye in matters of literary taste, they remained friends throughout the postwar decades. (For Mitchell, the relationship was irreparably tainted after Whitman's death by the publication of certain unflattering assessments of Mitchell's poetry in Horace Traubel's *With Walt Whitman at Camden.*)[61] Whitman also received medical treatment from Weir Mitchell's son, John K. Mitchell, under the direction of William Osler.[62] Osler wrote that he "sometimes went to Camden with Dr. J. K. Mitchell, always taking a greeting of some sort such as a book from Dr. Weir Mitchell."[63]

Whitman's financial situation in Camden was precarious. As his health deteriorated he increasingly relied on the support of friends and admirers. Mitchell was in a position to assist in various ways. Neither Silas Weir Mitchell nor his son John ever billed Whitman for their medical services.[64] In November 1889, Weir Mitchell wrote to Horace Traubel (Whitman's literary assistant and devotee during the Camden years) to discuss hiring a nurse for the aging poet. He closed by stating, simply, "It is needless to say that you may count on me."[65] When Traubel read Whitman the letter, he replied, "That is like the doctor—I know him and know of him—his goodness."[66] Mitchell paid fifteen dollars per month for the next two years to help cover the nursing costs. The renowned neurologist, who was also a prominent literary figure in his own right, regularly contributed additional funds toward Whitman's expenses.[67]

A More Wretched Spectacle

Mitchell entered the Civil War medical corps as a contract surgeon in October 1862, initially operating out of the old armory building at Sixteenth and Filbert Streets in Philadelphia. Unlike many of his colleagues, who were primarily focused on surgical anatomy, Mitchell was uniquely interested in injuries to the nerves. Prior to the war, he conducted vivisection experiments at the Philadelphia School of Anatomy. Defending the necessity of vivisection against accusations of cruelty, he insisted that cadavers alone were not enough to advance anatomical progress: "If we are to study successfully the workshops and factories of the body, we must study them while active and alive. . . . As it deals with life, it demands the aid of vivisection at every step. The knife of the operator lays bare the living, active organ, and enables us . . . to carry the torch of analytic chemistry into the midst of an organism still throbbing with vitality."[68] Civil War battlefields overflowed with human bodies to replace animals as subjects of scientific scrutiny, young men wounded but "still throbbing with vitality," whose collective suffering would transform American medicine.

As casualties flooded into Philadelphia, the armory hospital proved inadequate. In December 1862 Surgeon General William Hammond created a 400-bed hospital at a suburban estate on Turner's Lane, and placed Mitchell in charge of an entire ward dedicated to nerve disorders. As a result of this clinical experience, Mitchell published *Gunshot Wounds and Other Injuries of Nerves* (1864), which soon became the authoritative text on nerve injuries and treatment methods such as blistering, leeches, electricity, and cautery. Mitchell's intricate narrative is unusual among contemporary medical texts, revealing soldiers' names and biographical details in studies reminiscent of Whitman's "specimen cases." The following case history illustrates Mitchell's evocative storytelling. Far from impassive clinical rhetoric, Mitchell details Private S. Johnson's wounding and treatment in vivid prose:

> May 3d, 1863. — He was wounded by a small ball in the left cheek while riding at a trot. It entered at the middle of the ramus of the jaw . . . and finally lodged in the spinal column. . . . When shot, the man fell forward on his horse's neck; says he was confused, though conscious, and felt as if he had been struck in the ear, and then lifted up in air. He also felt instant pain in the back of his neck, and in all of his limbs. . . . He was removed from his horse, and carried to a house near by. . . . He now became aware of the total motor paralysis of the arms and legs. . . . Two days after being wounded, he became delirious, but gradually recovered his senses. . . . He was finally sent to Douglas Hospital, Washington . . . and was transferred to our own wards July 19th, 1863.[69]

Although the surface wound healed in nine weeks, damaged nerves continued to wreak havoc on Johnson's body, leaving his muscles atrophied and his flesh wasted. Mitchell's description of his first encounter with the soldier at Turner's Lane reflects the qualities that distinguish his voice across medical and fictional genres: striking visual imagery, lucid descriptions of pain, and dramatic narrative turns:

> July 19, 1863. Present state. — A more wretched spectacle than this man presents can hardly be imagined. He lies in bed, motionless, emaciated to the last degree, and with bed-sores on both elbows and both hips. His hands lie crossed on his chest, perfectly rigid; the fingers extended; the skin congested and thin; the nails curved . . . the head and neck rigid, with acute pain in these parts on movement. The right leg has motion of a feeble nature in all of the joints; the left only very slight voluntary movement. . . . These peculiarities have been modified by the long-continued rest of the limbs in one posture.[70]

The body's immobility, amplified by the limbs' "long-continued" rest, lends the soldier a statuesque quality. This "wretched spectacle" with arms resting across his chest evokes Christian iconography. Even the agnostic Whitman recalled the martyred Christ when viewing the corpse of an unknown soldier: "Young man, I think I know you—I think this is the face of the Christ himself, / Dead and divine and brother of all, and here again he lies."[71]

After artistically rendering his patient's condition, Mitchell turned his attention to documenting the medical treatment. He prescribed tonics and a high-caloric diet, alongside "shampooing and passive movement" of the soldier's fragile limbs. By August 25, Johnson's dermatological symptoms had abated, but the soldier was "still loosing flesh" and complained of "great pain in the neck." "Despite his prayers and protestations," Mitchell wrote, "the shampooing was continued; and, on October 15, he was greatly improved."[72] By January 1864, Johnson was well-nourished and regaining movement. Mitchell deemed him well enough to sustain a more aggressive therapy, and "ordered the patient to be etherized . . . and the limbs faradized daily."[73] A treatment favored by Mitchell in his postwar work on hysteria, faradization (also known as electrotherapy or electromagnetism) involved the application of an electrical current to stimulate the nerves. Building on Mitchell's research, faradization was often used as a treatment for shell shock during the First World War. Patients experiencing motor symptoms such as tremors, paralysis, contractions, limping, and fixed postures were repeatedly subjected to electrical currents. Faradization was heavily criticized in postwar Austria, to the degree that it was scarcely used at all during World War II. Wagner Jauregg, a Nobel laureate and professor of psychiatry in Vienna, was accused of cruelty in his practice of faradization and required to appear before an investigative committee that included Sigmund Freud.[74]

At least two of Whitman's physicians treated his postwar paralysis with faradization.[75] In a letter to Peter Doyle, Whitman wrote that Dr. Grier diagnosed his paralysis as the result of "the brain not being properly furnished and nourished with blood—(it is a disease the doctors call cerebral anemia)."[76] In his clinical notes on June 6, 1874, Dr. Grier describes the treatment that he administered to the poet: "descending secondary currents of Electro Magnetism to lower limbs. . . . For the brain—inverse constant current say about 6 to 10 Daniells cells from each sciatic N. to sacrolumbar region, not any higher under any circumstances. Every other day."[77] After years of chronic exhaustion and exposure in the wards, the poet who relentlessly documented the suffering of so many soldiers became a case study himself. Whitman endured experimental medical procedures in hopes of reversing his declining health. In a letter to his mother, the poet described his faith in electrotherapy, yet conceded that the treatment remained controversial: "I have had the second application of electricity to-day, quite a good application by Dr. Drinkard—he rubs the handles over my leg and thigh for about twenty minutes—the shock is very perceptible—it is not painful at all, feels something like pressing a sore—I feel as I did before, that it will be beneficial to me (though there are different opinions about this)."[78]

Whitman's description of faradization is all the more significant because Civil War-era medical narratives, even those as compelling as Mitchell's, often exclude patients' firsthand accounts of their sensory experiences in favor of the doctor's omniscient narration. We know, via Mitchell, that Johnson vehemently protested the shampooing of his limbs, and that his faradization was painful enough to require ether. Yet we do not know, in Johnson's own words, what either treatment *felt* like to him. Mitchell describes his methods as remarkably successful, despite their harrowing side effects. Following electrotherapy, Mitchell writes, Johnson's "progress was inconceivably rapid. . . . The patient was discharged in March, slightly shuffling in his walk, but with nearly entire use of all his limbs."

Replicas of Johnson's arms eventually found their way into the museum's cabinets: "Specimens Nos. 9 and 10, in the Army Medical Museum, are casts of both arms, and exhibit admirably their condition soon after they began to improve."[79] Mitchell was intimately acquainted with the museum, and not only because of his work with amputees. Brinton and Mitchell were both prominent members of the College of Physicians of Philadelphia, where they maintained close personal and professional alliances. Both were appointed to their Civil War posts by Surgeon General William Hammond, whom each considered a close friend. Two years later, the Army Medical Museum made a bizarre cameo in Mitchell's fictional debut of "spirit members."[80] Phantom limbs defied medical categorization for so long in no small part because the phenomenon resided entirely in the patient's sensations, as related to an external other.[81] The limb, being invisible, was also immeasurable. This inchoate pathology was pivotal to Mitchell's writings on the subject, both fictional and medical.

The Unseen Ghosts of Lost Parts

Throughout the war, and for decades after its conclusion, Silas Weir Mitchell collected data from amputees who continued to experience sensations associated with their severed limbs.[82] As late as 1892, Weir Mitchell and his son John distributed placards at the Army Medical Museum, inviting amputees to participate in a "study of the conditions of men who have lost limbs some time since."[83] As John wrote in the letter that accompanied the survey, "The matter is one which has never been investigated, and the only extensive material which exists for its study is among those who were unfortunate enough to lose limbs in the service of their country."[84]

Perhaps the strangest account of the Army Medical Museum occurs in Mitchell's fiction. In 1866, five years before his medical debut of the phantom limb, Mitchell adopted the pseudonym George Dedlow, a quadruple amputee afflicted by recurrent pain in his "lost members."[85] Framing amputation as a medical ghost story, "The Case of George Dedlow" was published in the *Atlantic Monthly* under the guise of an autobiographical essay. The story follows Dedlow from his placement as an army surgeon through the loss of each of his limbs. Following the final amputation, he awakens to find himself "more like some strange larval creature than anything of a human shape."[86]

Pseudonymity allowed Mitchell to explore the psychological facets of his research, outside the confines of scholarly discourse. Despite his suspicion that the inclusion of "psychical deductions" would cause some readers to dismiss his "metaphysical discoveries," Mitchell relied on the language of haunting to describe absent though active body parts for years to come. Mitchell hypothesized that Dedlow's severed nerves induced a form of anatomical nostalgia: "This pain keeps the brain ever mindful of the missing part, and preserves to the man a consciousness of possessing that which he has not."[87] Phantom pain acts as a memento mori, conserving perpetual awareness of the lost object. Each dismemberment not only erases another physical extremity, but further erodes Dedlow's subjectivity.[88] Mitchell describes this escalating apathy as a "deficiency in the egoistic sentiment of individuality."[89] Increased psychic detachment follows the severance of each limb, prompting Dedlow to question what remains of the body estranged from its "members":[90]

> One half of me was absent or functionally dead. This set me to thinking how much a man might loose and yet live. . . . I thus reached the conclusion that a man is not his brain, or any one part of it, but all of his economy and that to loose any part must lessen this sense of his own existence.[91]

Dedlow's existential disorientation echoes Mitchell's own deliberations on the "metaphysical" consequences of amputation.[92] As well as ghostwriting Mitchell's neurological theories, Dedlow also ventriloquized his patients' symptoms. The sensory resonance of the missing part—now "absent or functionally dead"—contrasts with an uncanny sense of wholeness. Soldiers who completed Mitchell's surveys as many as thirty years after their amputations often used the word "whole" to describe their phantom limbs. Responding to the question, "How much of the limb do you feel now, and how does the feeling differ from what it would be if the member were present," soldiers echoed a familiar refrain: "I feel the whole of both limbs"; "[it] feels just the same as if the whole limb was there."[93] One soldier captured this dichotomy in poetic simplicity: "I feel the whole of what is lost." Foreshadowing Mitchell's later research on sleep and sensation, a soldier whose left arm was severed at the elbow reported feeling "the entire [amputated] arm as though normal." However, the intact "arm and hand feel as though asleep and shrunken." The phantom's sensory normality eclipses the amputee's altered perception of his remaining arm; the body is less real than its ghost. Like Whitman's case histories of Brown and Geiger, several of the soldiers Mitchell surveyed reported entanglements and contractions of their lost fingers and toes.[94]

Mitchell's interrogation of the amputee's vanishing "economy" recalls Brinton's account, detailed in chapter 2, of the soldier who attempted to reclaim his limb from the museum, in spite of the curator's insistence that the Union retained control over the soldier's body, and any detached parts, for the duration of the war. Dedlow's meditation on the psychological consequences of amputation also parallels Whitman's attention to the wound's magnetic vacancy. Just as Whitman nostalgically returns to the hospital wards, Dedlow is so compulsively drawn to his lost limbs that he conjures them at a spiritualist séance, where the legs announce their arrival by spirit-rapping their specimen numbers in the Army Medical Museum.[95] Dedlow is then able to stand and walk a few brief steps on "limbs invisible."[96] This supernatural reunion is inevitably fleeting. Following the séance, Dedlow remains a "fraction of a man," longing for "the day when I shall rejoin the lost members of my corporeal family in another and happier world."[97] Dedlow's kinship with his lost limbs reflects changing cultural narratives of corporeal possession in the wake of the war's carnage. While the state may retain ownership of the physical specimen, the amputee's attachment to his "spirit member" cannot be severed so easily.[98]

Civil War amputees experienced dissonant responses to encounters with their lost parts, whether physical or spectral. In contrast to those who suffered anxiety about their limbs' public preservation, Union General Daniel Edgar Sickles donated his leg to the Army Medical Museum and made an annual pilgrimage to visit the specimen on the anniversary of his amputation. Brinton recalled this anecdote in an 1896 address to the Army Medical School: "Occasionally it would happen that a specimen would be contributed to the museum by its former owner. No. 1335, resulting from a leg crushed by a twelve-pound shot at Gettysburg, and for which a thigh amputation was performed, was forwarded to the museum in an extemporized coffin on which was tacked a visiting card 'with the compliments of Major General D.E.S., United States Volunteers.' But not all were so compliant."[99] Brinton's disregard for "[un]compliant" soldiers links the museum's curatorial practice with medical body snatching. Disempowered subjects make better specimens (except for the occasional donation from a notable figure), as they have little agency to resist anatomical appropriation. Recalling Johnson's complaints about Mitchell's treatment of his limbs, one cannot help but wonder if the soldier consented (or was even asked) to have casts of his arms displayed in the museum. In these cases, the question of medical consent correlates with military rank and social status. Like Johnson, the soldier that Brinton describes demanding the return of his limb was a private, whereas General Sickles revels in his specimen's public display. Such varying hierarchical reactions demonstrate the privilege afforded to medical and military agencies to control the postmortem gaze, with or without consent. The state and the surgeon wielded the authority to retain possession of human bodies and their parts, long after the "original possessor" turned to dust.[100]

No Resurrection of the Dead

In late 1866 donations began flooding into Philadelphia's South Street Hospital, commonly known as "Stump Hospital." Scores of visitors arrived to pay their respects to George Dedlow. So compelling was Mitchell's tale that many callers refused to believe that no such patient had ever resided there.[101] Mitchell felt compelled to offer a veiled apology for this unintentional deception in the article that finally named the phenomenon, "Phantom Limbs," published in *Lippin-*

cott's Magazine of Popular Literature and Science (1871).[102] A vehement debunker of occultism, Mitchell opened his inaugural medical analysis by dismissing Dedlow's séance as absurd.[103] Yet, the specter is not so easily exorcised. Like the amputees he observed, Mitchell's medical prose is haunted by "ghostly members, which seem so material."[104] Even the name given to the "neuralgic torture to which stumps are liable" applies spiritualist rhetoric to post-amputation sensations: "thousands of spirit limbs haunting as many good soldiers . . . tormenting them with the disappointments which arise when, the memory being off guard for a moment, the keen sense of the limb's presence betrays the man into some effort, the failure of which suddenly reminds him of his loss."[105] The lapse between cognitive logic and the limb's false presence recalls Dedlow's assertion that "the pain keeps the brain ever mindful of the missing part"—though Mitchell's theory expanded beyond pain to include a host of sensations that must be guarded against. Describing the return of a phantom to a soldier who had long ceased to feel it, catalyzed by electric shock, Mitchell again likened the effect to the appearance of an *invited* apparition: "no resurrection of the dead, no answers of a summoned spirit, could have been more startling."[106] Mitchell observed that phantoms are often fragmented or misshapen, gradually eroding until the lost hand is felt at the elbow, or the foot at the knee: "while most men are conscious of a lost limb as still in place, the spirit member is never complete."[107] Similar to Freudian melancholia, the phantom limb is attached to an infinite wound; the "spirit member" is permanent only in its partiality. Like Lewy Brown's entangled toes, Mitchell found that phantom extremities were often frozen in contortion, or replicated the posture just prior to removal.[108] Mitchell echoes Whitman's account of Thomas Geiger's flexing fingers almost verbatim: "sometimes the ghostly members are in a perpetual state of automatic activity, and the fingers open and shut or twist across one another."[109]

Mitchell's neurological treatise proved prophetically accurate; many of the symptoms he discovered have become accepted features of phantom limb syndrome. Ronald Melzack, whose pioneering work revolutionized pain research, identified two fundamental characteristics that reaffirm Mitchell's original thesis—reality and belonging:

> The most extraordinary feature of phantoms is *their reality to the amputee*. Their vivid sensory qualities and precise location in space—especially at first—make the limbs seem so lifelike that a patient may try to step off a bed onto a phantom foot or lift a cup with a phantom hand. The phantom, in fact, may seem more substantial than an actual

limb, particularly if it hurts. . . . A final striking feature of phantoms, which reinforces the reality still further, is that they are experienced as a *part of oneself*. That is, patients perceive them as integral parts of the body. A phantom foot is described not only as real but as unquestionably belonging to the person.[110]

Writing over a century apart, Melzack and Mitchell agree that the phantom limb manifests as a "material" presence, which often feels more "substantial" than the intact body.[111] Enhancing this tangibility is a lasting sense of ownership over the lost part. Melzack's description of the phantom as "unquestionably *belonging* to a person" once again recalls the antebellum controversy surrounding bodies as property. While Brinton argues in favor of perpetual military and medical appropriation of human parts, Mitchell's phantom limbs disturb this logic. The specimen may be detached, preserved, and displayed but the amputee retains its ghost. This haunting subverts time and space, corporeality and authority, insisting that the absent part is not entirely lost. Tellingly, Mitchell's most evocative example of post-amputation amnesia again references the Army Medical Museum. Mitchell discusses several case histories in "Phantom Limbs," but only twice does he quote his patients directly. One account narrates a soldier's recurrent fear upon waking that his limb has become a specimen: "every morning I have to learn anew that my leg is enriching a Virginia wheat crop or ornamenting some horrible museum."[112] In the following decades, Mitchell's research increasingly focused on links between sleep, memory, and sensation. As he collected data from amputees, Mitchell discovered that dream phantoms are as varied and material as those that haunt waking life.

Phenomenologically, the phantom limb manifests as a physical presence felt most acutely in its absence. Mitchell observed that amputees "seem to retain a sense of its existence so vivid as to be more definite and intrusive than is that of its truly living fellow-member."[113] This neurological nostalgia parallels Whitman's melancholic drive to textually preserve specimens. As Moon observes, Whitman narrates "desire(s) of or for bodies that are no longer capable of being perceived as whole, healthy, and labile."[114] The phantom experience is no longer pathologized as a neural aberration. Recalling Mitchell's query into "how much a man might lose and yet live," Peter Halligan observes that the medical denigration of phantoms was "based on a long-standing and pernicious folk assumption that the physical body is necessary for experience of a body."[115] Whitman's specimens bridge the space between ghost and host. A vehicle for enduring cathexis, the specimen signifies an erotics of partiality—mourning an absence that is both embodied and spiritual. For Whitman, the physical body is not necessary to sustain attraction to the lost other, or to the part of one's self that now resides with the dead. This reverence toward partial bodies demonstrates an attachment to the process of loss, through which profound intimacies are formed.

Whitman's libidinal investment in amputees mirrors his fascination with erotic-linguistic vacancy. *Memoranda*'s specimen "cases" borrow from contemporary botanical and medical rhetoric to chart the evolution of a unique category of "beings."[116] As Roper explains, "Walt and other men and women of the last half of the nineteenth century were aware of a lack of terminology for describing a way of being they felt themselves, sometimes, in whole or in part, to embody."[117] The term "homosexual" first appeared in English in a translation of Richard von Krafft-Ebing's *Psychopathia Sexualis* in 1892, the year of Whitman's death. The poet's appropriation of words such as phrenological "adhesiveness" for masculine same-sex desire, "specimen" for subject of erotic curiosity, and "comrade" for lover or friend, are attempts to construct a nomenclature which could fill that void. *Memoranda*'s collected observances represent an epistemological study in keeping with Foucault's historicity of homosexuality: "The nineteenth-century homosexual became a personage, a past, a *case history*, and a childhood, in addition to being a type of life, a life form, and a morphology, with an indiscreet anatomy and possibly a mysterious physiology."[118] Whitman offers a unique perspective on the evolution of this anatomy: he views queer morphology through reverential, rather than diagnostic, eyes. In the context of his hospital work, this sense of queer unspeakability operates at the threshold of contemporary medical investigations.

Just as the phantom limb preserves a sensory link to the lost part, Whitman experienced nostalgia for the hospitals:

> Thus in silence, in dreams' projections,
> Returning, resuming, I thread my way through the hospitals,
> The hurt and wounded I pacify with soothing hand,
> I sit by the restless all the dark night, some are so young,
> Some suffer so much, I recall the experience sweet and sad,
> (Many a soldier's loving arms about this neck have
>   cross'd and rested,
> Many a soldier's kiss dwells on these bearded lips).[119]

Like the ghostly pains of the amputee, Whitman inevitably returns, "in dreams' projections," to the hospital corridors. The poet seeks to psychically resurrect soldiers' abandoned bodies and detached parts. The phantom limb is literally an untouchable extension, a neurological trace of the lost object. This elegiac rendering of the deathbed scene evokes a similarly impossible touch: the revivification of a dying man's kiss.

The reciprocal fidelity of convalescence rejects hetero-normative assumptions that sexuality is negated by partiality. Whitman highlights the loyalty of the wounded toward their "dresser": "I have come to adapt myself to each emergency . . . washing and dressing the wounds (I have some cases where the patient is unwilling any one should do this but me)."[120] For the self-proclaimed "poet of the body," the wound occupies a space of heightened sensation, inhabiting the borderland between interior and exterior, mortality and spirit.[121] Sedgwick elucidates a theoretical paradox around such corporeal enclaves of vacancy: "erotic localization has the effect of voiding—of voiding by so exceeding it—the very possibility of erotic localization. . . . The neat dichotomy of 'active' and 'passive' renders as an organ of this sexuality the whole cutaneous envelope of the body."[122] Whitman figures the wound as a space of heightened immersion, home to an eroticism that transcends physicality by spectacularly eclipsing it. Within this threshold, suffering acts as a portal for "animal purity," catalyzing the transition from flesh to spirit:

> *Spiritual Characters Among the Soldiers.*—Every now and then in Hospital or Camp, there are beings I meet—specimens of unworldliness, disinterestedness and animal purity and heroism . . . the power of a strange, spiritual sweetness, fibre and inward health have also attended. Something veil'd and abstracted is often a part of the manners of these beings. I have met them, I say, not seldom in the Army, in Camp, and in the great Hospitals. . . . They are often young men . . . unaware of their own nature (as to that, who is aware of his own nature?), their companions only understanding that they are different from the rest, more silent, "something odd about them," and apt to go off and meditate and muse in solitude.[123]

Whitman's specimens exist within a class all their own, members of some higher order of "beings." There is something already spectral, "veiled and abstracted," about their presence, which haunts the poet in advance of their actual deaths. The specimen epitomizes the intimacy that passed between strangers in the hospital wards. Oddness is an integral aspect of his character, an uncanny articulation of queer specificities that render the subject "unworldl[y]." This strange entity not only resists mortality, his presence somehow transcends the material world.

Like Mitchell's "spirit members," Whitman's description of soldiers as "spiritual characters" situates his specimens as performers in a fashionably macabre contemporary discourse. Spiritualists embraced modern innovations in science and technology as a means to their particular end. Spirit photographs were a popular form of memento mori intended to capture "portraits of psychical entities not seen by normal vision."[124] Spirit rapping, an auditory method of communicating with the deceased, was referred to as the "Spiritualist telegraph," while Alexander Graham Bell's assistant, Thomas Watson, later "imagined the telephone as an aid to spiritual connection."[125] As early as 1855, Whitman associated poetry with spiritualism, casting writer and medium as magnetic conduits: "The poets are divine mediums—through them come spirits and materials to all the people, men and women."[126] As liminal occupants of both corporal and ethereal realms, the specimen offers a queer alternative to mourning practices dominated by elaborate funerary traditions. Its mourners do not require a body, or even part of a body, to situate their loss. Through the textual preservation of this ephemeral being, infinite others can be absorbed.

While the war shattered Whitman's vision of a cohesive Union, its soldiers embodied his "Calamus" ideal of "the manly love of comrades."[127] The plant was traditionally used in naturopathy as a balm for "slowly healing wounds."[128] In *Botanologia* (1710) William Salmon recommended calamus as a cataplasm: "applied to the Testicles, it wonderfully abates their Swelling."[129] Whitman would have been aware of these medicinal properties. As Maria Farland has demonstrated, "Whitman's medical philosophy was deeply rooted in a set of ideas linked to naturopathic and herbal healing."[130] The poet applies his "Calamus" principles of "adhesiveness" as a curative salve, an act that is highly subversive and erotically charged:

> I dress a wound in the side, deep, deep;
> But a day or two more—for see, the frame all wasted and sinking,
> And the yellow-blue countenance see.
> . . .
> I am faithful, I do not give out,
> The fractur'd thigh, the knee, the wound in the abdomen,
> These and more I dress with impassive hand (yet deep in my
>   breast a fire, a burning flame).[131]

The "burning flame" in Whitman's breast recalls two other mysterious medical conditions that originated at Turner's Lane Hospital. Causalgia, also discovered by Mitchell, refers to a burning pain in the limbs caused by a peripheral nerve injury.[132] This pain was so severe that Mitchell recalled, "we have again and again been urged by patients to amputate the suffering limb."[133] The phenomenon known as "soldier's heart" was given its name by Jacob Mendes Da Costa, who supervised the ward adjacent to Mitchell's, where he documented enigmatic cardiac disturbances: "I noticed cases of a peculiar form of functional disorder of the heart."[134] Symptoms of "soldier's heart" included cardiac pain, palpitations, shortness of breath, sleep disruptions, and digestive complaints—all now understood to be psychosomatic manifestations of trauma. Da Costa and his colleagues searched in vain for a physical cause that could explain their patients' suffering. Whitman described the dichotomy between physical wounding and heartache in an earlier notebook: "you break your arm and a good surgeon sets it and cures you complete; but no cure ever avails for an organic disease of the heart."[135]

Many of Whitman's war texts are meditations on the pleasures of tending unhealing wounds, recalling Freud's assertion that melancholia "behaves like an open wound" that seeks to fill itself entirely with absence.[136] Writing to a friend in New York to explain his relocation to Washington, Whitman described the perpetual adhesion that bound him to the wounded:

> The amputated, sick, sometimes dying soldiers cling & cleave to me as it were as a man overboard to a plank, & the perfect content they have if I will remain with them, sit on the side of the cot awhile, some youngsters often, & caress them &c.—It is delicious to be the object of so much love & reliance, & to do them such good, soothe & pacify torments of wounds.[137]

The soldier's clinging is akin to the grasp of a drowning man, pacified by the poet's caress. Their mutual "reliance" embodies a melancholic attachment forged in trauma, outlasting death. Although he laments the "torments of wounds," Whitman also finds it "delicious to be the object of so much love." Despite the soldier's suffering, from Whitman's perspective there is reciprocal pleasure (however fleeting) in this tactile exchange between the sufferer and his attendant. Like the absent presence of the phantom limb, the traumatic vacancy of melancholia can never be filled. This interior lesion remains not only open, but insatiable: a "foreign body" embedded within the psyche that continually attracts libidinal investment.[138] The melancholic refuses to divest attachment from the absent other, denying the act of psychic prosthesis that would sever cathexis with the dead.[139] Like the sensory ghosts of lost limbs, Whitman's words preserve a living, "breathing" connection to "countless" lost soldiers.[140] In the war's diasporic aftermath, Whitman moves beyond the grave as the scene of "erotic localization."[141] The book becomes the tomb; the phantom eclipses the corpse as the poetic object of enduring attachment. To absorb the "million dead," *Leaves of Grass* is reconfigured as a nostalgic crypt in which infinite bodies are preserved, awaiting, upon the invocation of the poet-medium, their resurrection.[142]

What Deepest Remains

Civil War soldiers suffered from a range of maladies for which there was no corresponding physical wound. After insanity, nostalgia was the most common diagnosis employed to describe symptoms that would today be attributed to traumatic stress.[143] The disorder's military origins were evident in Johannes Hofer's 1688 thesis on a malignant strain of homesickness that afflicted Swiss soldiers. Hofer created the term "nostalgia" from the Greek words for "return" (*nosos*) and "sorrow" (*algos*) to describe a psychosomatic malady: "the continuous vibration of animal spirits through those fibers of the middle brain in which impressed traces of the Fatherland still cling."[144] Hofer's restless "animal spirits" foreshadow the "animal purity" of Whitman's specimens and the neurological ghosts of Mitchell's "fractional phantoms."[145] Preceding by more than two centuries Freud's theory of hysteria as excessive reminiscence, Hofer pathologized nostalgia as "disorder[ed]" memory—an unrelenting "desire" for the past.[146] The impossibility of return elicited a terminal melancholy that "exhausted" the "vital spirits" of the mind:

> Nostalgia is born from a disorder of the imagination. . . . The nostalgic are affected but by few external objects, and nothing surpasses the impression which the desire to return makes on them. . . . Melancholy plays a part here, for the vital spirits, worn out by the single idea which occupies them, become exhausted and provoke erroneous representations.[147]

At the time of Hofer's work, medical science was attempting to comprehensively inventory and classify disease, similar in practice to the botanical system. Hofer's project was unprecedented in its conversion of an emotional phenomenon into a medical diagnosis. He created a new category of disease, which merged the physical with the spiritual. From its inception as a disease of the *body*, nostalgia was embedded within the *psyche*: the physical manifestation of desire for an impossible return, a longing not for a specific location, but for an interior geography that is ephemeral in its very nature.

By the eighteenth century, nostalgia was firmly established as a medical disease with a specific set of diagnostic criteria, and theoretically a single cure: home.[148] On treatment and progression, Hofer was adamantly concise: "This ailment is curable if the yearning (*Sehnsucht*) can be satisfied; incurable, mortal, or at least very grave when circumstances prevent its satisfaction."[149] Yet, medical researchers in the following century found that fulfillment of the desire for return did not always alleviate suffering. Even after homecoming, this peculiar melancholy did not necessarily recede. In cases where the nostalgic was fundamentally altered during his absence, his sadness remained unabated.[150] Medical research increasingly focused on the phenomenon's military significance, linking nostalgia with suicide and desertion. The disorder was particularly virulent in conscripted soldiers from rural areas, for whom the memory of home became something akin to a pastoral ideal. Philippe Pinel's entry on nostalgia in *Encyclopédie Méthodique* (1821) linked the disorder's progression with prolonged stays in hospital, which in turn increased the subject's likelihood of contracting other diseases: "this extended stay is almost always mortal, for sooner or later they are stricken by the diseases which permeate, in a frightful way, the military hospitals, such as dysentery, and fevers accompanied with loss of strength."[151] The secondary fatality of nostalgia—increased risk of contracting one of the many "diseases which permeate[d]" military hospitals—parallels Whitman's perception of his illness as the result of ingesting the "poison" that "saturated" the wards.[152]

Although nostalgia had declined as a diagnostic category by the mid-nineteenth century, the Civil War catalyzed its revival.[153] The Union Army's *Manual of Instructions for Enlisting and Discharging Soldiers* (1863) defined nostalgia as a "mental disease" within "the class of Melancholia," characterized by an "unconquerable longing for home" that often proved fatal.[154] Symptoms included "loss of appetite," "hectic fever," "a dull pain in the head," "throbbing of the temporal arteries," "anxious expression of the face," "watchfulness," and finally "a general wasting of all the vital powers."[155] The official records of noninfectious diseases disclose 5,213 cases and 58 deaths from nostalgia among white Union troops from May 1861 to June 30, 1866, and 334 cases and 16 deaths among African American soldiers over the same period.[156] Medical officers suspected that nostalgics were affected by the "poisonous effluvia" of overcrowded camps; men from rural areas were believed to be more susceptible than city-dwellers, who had developed a resistance to "crowd poisoning."[157] Due to the disorder's high mortality rate, it was grounds for medical discharge. Speaking to the Physicians Club of Chicago in 1913, Weir Mitchell described the severity of the epidemic:

> Cases of nostalgia, homesickness, were serious additions to the peril of wounds and disease, and a disorder we rarely see nowadays. I regret that no careful study was made of what was in some instances an interesting psychic malady, making men hysteric and incurable except by discharge. Today, aided by German perplexities, we would ask the victim a hundred and twenty-one questions, consult their subconscious mind and their dreams, as to why they wanted to go home and do no better than let them go as hopeless.[158]

Mitchell's disdain for the burgeoning field of psychology is palpable. While Freud described his "talking cure" as an evolution of Mitchell's "rest cure" and favorably reviewed several of the neurologist's writings, Mitchell repeatedly derided psychoanalysis.[159] Paradoxically, Mitchell's research involved the very methodology that he dismisses in the above address. In the postbellum decades, he distributed detailed surveys containing dozens of questions to amputees in an effort to understand their invisible ailments. He later became fascinated by the relationship between sleep and sensation, which led to his research on the phenomenology of dreams. Like Mitchell's "metaphysical discover[y]" of phantom limb, nostalgia wreaked havoc on body and mind, altering sufferers' perceptions of time, space, and somatic experience. Forty-one years after the war, Mitchell received the following letter from the veteran H. S. Huidekoper. I have come to think of this prominent Philadelphian, awarded the Medal of Honor for his Civil War service, as the "ghost writer":

> As with everybody else who has lost a limb, the fingers are distinctly felt, and pains occur oftentimes to various parts. . . . Now for the curious part. I was 24 years old when I lost my arm, and am now 67. Almost two-thirds of my life has passed without thought of the possible use of my right arm, and yet never have I dreamed once, that I was without two arms . . . I write often in my dreams, but always with the right hand I used over forty years ago. To do this, I attempt to use the tendons which would hold and guide the pen, and this is done with so much fatigue . . . that

> I suffer great pain in my finger tendons, even to waking me up from the most profound sleep, because of the pain in the lost hand. Thus, in my dreams, I remain a man with a perfect frame, but while awake, I never think of myself as otherwise than a one-handed being.[160]

Perhaps inspired by Huidekoper, five years later Mitchell began investigating dream consciousness. A letter from the physician William Sturgis Bigelow (March 8, 1911) responds to the neurologist's questions as to whether his "dreams always [had] a relation to events of waking life" and if he had in dreams "the power to see things visualized . . . to occupy a part of the predormitum with such visions as I chose to call up?"[161] Soliciting responses from both medical colleagues and literary scholars, Mitchell examined the sleeper's agency over his dreams: can one initiate lucid dreams by visualizing certain objects or scenes in the period before falling asleep?[162] Could dreams be, in Freud's words, "a wish fulfilled?" Indeed, Mitchell's queries parallel Freud's interrogation of the unconscious in *The Interpretation of Dreams* (1900): "What changes have occurred in the dream thoughts before they are transformed into the manifest dream which we remember upon awaking? In what manner has this transformation taken place? Whence comes the material which has been worked over into the dream?"[163] Just as he relied on spiritualist terminology despite his cynicism toward haunting when writing about phantom limbs, Mitchell's inquiry into the subject's influence over his dream life bears striking correlations with psychoanalysis. Mitchell's fascination with dreams also foreshadowed future research on phantom limbs. During the twentieth century, dream states became integral to neurologists analyzing the phantasmal body scheme, offering a lens into how dismembered ghosts intersected with the physical body. As Crawford explains, dreams "revealed that even the most apparently well-adjusted amputees wished for the kind of physical integrity that wholeness offered . . . As unrealistic as these idyllic dreams—or, for some, agonizing nightmares—were, they purportedly revealed a universal desire or need to deny the loss of a limb, a need for physiologic intactness."[164]

Huidekoper's recurrent nightmare of aborted writing recalls the phantasmal soldiers who arise from the pages of Whitman's books. For both men, the act of writing is an invocation. While Whitman returns to the hospitals "in dream's projections," Huidekoper reunites with his lost writing hand in sleep. Whitman's ghosts reside in pages of his "bloodstain[ed]" notebooks, while Huidekoper is woken by phantom pain as he grips the pen to write in dreams. Almost two decades after the war, Whitman's account of somnambulist nostalgia, "Old War Dreams" (1881, 1892), demonstrates the adhesiveness of trauma through the subject's return to the past in "midnight sleep":

> In midnight sleep of many a face of anguish,
> Of the look at first of the mortally wounded (of that indescribable
>   look)
> Of the dead on their backs with arms extended wide,
> I dream, I dream, I dream.
>
> Of scenes of Nature, fields and mountains,
> Of skies so beauteous after a storm, and at night the moon so
>   unearthly bright,
> Shining sweetly, shining down, where we dig the trenches and
>   gather the heaps,
> I dream, I dream, I dream.
>
> Long have they pass'd, faces and trenches and fields,
> Where through the carnage I moved with a callous composure,
>   or away from the fallen,
> Onward I sped at the time—but now of their forms at night,
> I dream, I dream, I dream.[165]

The opening stanza recalls the burial squads at Fredericksburg, recorded in Whitman's notebook on his first visit to the front. An "unearthly" moon shines "sweetly" over open trench graves, creating an uncanny dissonance between the "beauteous" nocturnal landscape and the "carnage" of war. The poem's speaker retains a medical objectivity more in keeping with a military surgeon than the hyper-attachment Whitman narrates in *Memoranda* and his private notebooks. Despite his "callous composure . . . at the time," the dead return years later to haunt his dreams.

The impossibility of nostalgic return was never more historically evident than during the Civil War. Soldiers who lived to return home found the once-familiar landscape as ravaged as their own psyches. Amputees were a visible reminder of the war's carnage, evoking both fascination and anxiety. The unseen wounds of trauma bore other, internal scars. Whitman's hospital service had lasting psychological and physiological consequences. As his health steadily declined throughout the postbellum decades, the poet considered himself a casualty of war, alongside the soldiers he had nursed. While Whitman's primary wartime illness was tubercular in origin, the poet suspected that the cause of his waning health was not entirely physical.[166] He described his deterioration as "tenacious, peculiar, and somewhat baffling."[167] In a letter to Lewy Brown, Whitman narrated the onset of his illness.

> My dear comrade I have been very sick, and have been brought on home nearly three weeks ago, after being sick some ten days in Washington—The doctors say my sickness is from having too deeply imbibed poison into my system from the hospitals—I had spells of deathly faintness & the disease also attacked my head & throat pretty seriously. The doctors forbid me going any more into the hospitals—I did not think much of it till I got pretty weak and then they directed me to leave and go north for change of air as soon as I had strength. . . . This is the first sickness I have ever had & I find upon trial such things as faintness, headache & trembling and tossing all night, & day too, are not proper companions for a good union man like myself.[168]

Whitman's diagnosis of "imbibed poison" reflects nineteenth-century conceptions of disease as the consequence of absorbing toxic miasmas. Whitman's "hospital fatigue" was also remarkably similar to Assistant Surgeon General Dewitt C. Peters's description of nostalgia as a "species of melancholy" (1863).[169] Many of the symptoms that triggered Whitman's collapse were also associated with chronic nostalgia: depression, insomnia, night terrors, dizziness, fainting, headaches, photophobia, sore throat, and a recurrent humming in the ears.[170] Speaking to Horace Traubel many years later, Whitman recalled the war's sensory impacts: "Some days I was more emotional than others, then I would suffer all the extra horrors of my experience—I would try to write, blind, blind, with my own tears."[171] The "horrors" of war returned irrepressibly, recorded in Whitman's numerous accounts of flashbacks, nightmares, and spectral visions: "Even here in my room-shadows and half-lights in the noiseless flickering flames, again I see the stalwart ranks on-filing, rising—I hear the rhythmic tramp of the armies."[172]

Whitman repeatedly unsettles the temporality of traumatic memory. His war recurs outside of time; hallucinations are induced by reading, writing, or sleeping. He articulates the sudden onset of flashbacks and their physical manifestations: "When I am home or out walking alone, I feel sick and actually tremble when I recall the thing and have it in my mind again before me."[173] In keeping with Hofer's depiction of nostalgia as disordered memory, Paul Ricoeur characterized the "obsessional eruptions" that follow trauma as symptoms of "wounded or sick memory."[174] Reflecting on the psychosomatic elements of recovery, Whitman's words to Traubel echo the classification of nostalgia as a "species of melancholy": "Any Doctor will tell you how necessary it is—a species of mind cure."[175]

Just as the phantom limb exists outside temporal and physical borders, so too does trauma. In "The Wound-Dresser," Whitman describes the psychic intrusions of combat violence: "What stays with you latest and deepest? of curious panics, of hard-fought engagements or sieges tremendous what deepest remains?"[176] The question of "deepest remains" alludes to the "unfound" dead and those who were "hastily buried" in shallow graves.[177] For surviving soldiers, the traumas of war were likewise only partially interred, able to resurface at any moment in "curious panics." This play on words juxtaposes "deep" and enduring psychic reverberations with the shallowness and impermanence of field graves. Unrecovered (and uncovered) human remains are, paradoxically, the "deepest remains" of the wounded psyche, they "stay latest" and longest, resurrected in flashbacks of the sieges that claimed their lives.

For the traumatized individual, consciousness is hostage to an uncontrollable immediacy. As Jane Goodall and Christopher Lee observe in their introduction to *Trauma and Public Memory*, the "past is recalled to the present with an urgency that violates temporality as a structuring principle of mental and emotional life."[178] The epitome of this somatic time-travel is the flashback experience, when physiological sensations flood the subject, defying linear logic. Physical responses to triggering events (reminiscent of Whitman's symptoms) include tremors, nausea, vertigo, palpitations, paralysis, and hallucinations. The hallucinations evoked by flashbacks—visual, auditory, and olfactory—are not mere delusions, but rather invasions of the subject's very real past into the present moment. Like phantom limbs, these sensory resurrections reflect acutely embodied perceptions; trauma ruptures the subject's connection to "the time-scheme of a shared world."[179]

In "A March in the Ranks Hard-Press'd and the Road Unknown," Whitman narrates a hallucinatory return to the field hospital at Lacy House:[180]

Faces, varieties, postures beyond description, most in
    obscurity, some of them dead;
Surgeons operating, attendants holding lights, the smell
    of ether, the odor of blood;

The crowd, O the crowd of the bloody forms of soldiers
—the yard outside also fill'd;
Some on the bare ground, some on planks or stretchers,
some in the death-spasm sweating;
An occasional scream or cry, the doctor's shouted orders
or calls;
The glisten of the little steel instruments catching the
glint of the torches;
These I resume as I chant—I see again the forms, I
smell the odor;[181]

The resurrected crowds of soldiers and surgeons and the accompanying scent of ether and blood seem at first glance to be a seamless rendering of the flashback experience. Yet there is a remarkable distinction: these multisensory hallucinations are not the uncontrollable consequence of a trigger. The poet's chant calls them into being. Challenging the theory that traumatic memories induce a terrifyingly passive state over which the subject has no agency, Whitman imagines the phenomena we currently term post-traumatic stress not as flashback but as visitation. While Mitchell seeks to determine whether the sleeper has agency over his dreams, Whitman returns to the traumatized subject's agency over the waking nightmare of the flashback. In "Ashes of Soldiers," he invites the infinite companionship of spectral bodies:

Ashes of soldiers South or North,
As I muse retrospective murmuring a chant in thought,
The war resumes, again to my sense your shapes,
And again the advance of the armies.

Noiseless as mists and vapors,
From their graves in the trenches ascending,
From cemeteries all through Virginia and Tennessee,
From every point of the compass out of the countless graves,
In wafted clouds, in myriads large, or squads of twos or threes or
single ones they come,
And silently gather round me.

Admitting around me comrades close unseen by the rest and
voiceless,

The slain elate and alive again, the dust and debris alive,
I chant this chant of my silent soul in the name of all dead
  soldiers.
Faces so pale with wondrous eyes, very dear, gather closer yet,
Draw close, but speak not.

Phantoms of countless lost,
Invisible to the rest henceforth become my companions,
Follow me ever—desert me not while I live.[182]

The war "resumes" in the poet's mind, a "retrospective" haunting conjured by internal "murmuring." Like Mitchell's "sensory ghosts," Whitman's phantoms are "unseen" and unheard—but not unfelt: "again to my *sense* your shapes." Just as amputees continue to embody the "unseen ghosts of lost parts," Whitman's slain soldiers are "alive" yet elusively partial as "dust and debris," intangible "as noiseless as mists and vapors." Compared to *Memoranda*'s heartbreakingly specific "specimen cases"—its catalog of each soldier's individual "character" alongside biographical and medical histories and evocative details such as a craving for preserves or ice cream—these phantoms are pale shades of their original specimens. In this sense the elegy fails to remember, erasing the names and identifying marks of its ghosts in order to render them more accessible to a nation of mourners. Yet, as Adam Bradford has argued, the phantom's anonymity is also a form of hospitality.[183] The unseen ghost is a prismatic figure who invites the projection of "countless" losses. The chant refuses to name these specters, visible only to the poet as medium, so that the reader might see instead the face s/he most longs for.

In keeping with Hofer's theory, Whitman's nostalgia desires an impossible return to the past. But it also entails an unending desire for that desire. Whitman longs to remember, no matter what the cost. Like the entangled nerves of a severed but still sensate limb, Whitman saw the war's ghosts as integral to *Leaves of Grass:* "My Book and the war are one / Merged into its spirit I and mine." The soldiers' phantoms are implored to remain by the poet's side. His memories may at times arrive unbidden, but he would never wish them away. As Buinicki writes, Whitman casts his Reconstruction-era poems as "sites of memory" that "reconstruct his experiences as they slip inexorably into the past."[184] Theoretically, the nostalgic longs for a revered past, while the traumatized subject returns incessantly to the chaos of suffering. Yet for Whitman, these scenes (the idealized past and the traumatic event) were one and the same: the "magnetic and terrible sights" of the war hospitals.[185] Just as Whitman derived paradoxical pleasure (alongside immeasurable grief) from tending soldiers' wounded bodies, the same is true of his own wounded mind. The flashback is welcomed, even invoked, for it recalls the lost specimens. In this sense, Derrida's ethos of reciprocal haunting echoes Whitman's literary conjurations: "an open hospitality to the guest as *ghost*, whom one holds, just as he holds us, hostage."[186] Call it nostalgic reverie or traumatic intrusion, if a sensory recurrence brought with it the lost other, would one not come to tolerate, even to invite, the flashback? No matter how unsettling, is there anything the bereaved longs for more than reunion with the dead? Even Mitchell's amputees, so long tormented by their limbs' phantoms, dreamed of reunification with their "lost members."[187] As Mitchell suspected, neurologists have come to acknowledge the liminality of phantoms—to recognize that the body can replicate its detached parts outside time and space. Likewise, Whitman offers a way of reading trauma's return and rupture pattern that is open to the ambiguity inherent in violent death. Recalling the word's etymology, the Whitmanian "specimen" operates as a kind of lens, something through which we might fleetingly behold "countless" lost others. For no matter how much pain the inevitable rupture brings, the mourner would endure all this and more for mere moments with the ghost.

4 Skeleton Leaves
Embalming Elegies

As casualties mounted in the deadliest conflict in American history, recovering battlefield fatalities became a logistical nightmare.[1] Due to the urgency of troop relocation and the obstruction of officers who refused access to enemy burial parties, thousands of corpses remained where they fell. As Union Major General George Meade lamented after his costly victory at Gettysburg, "I cannot delay to pick up the debris of the battlefield."[2] An entrepreneurial generation of photographers captured this "Harvest of Death," rendering these bodies forever unburied in the eyes of viewers.[3] As Max Cavitch writes, "gallery exhibitions, lavish albums, *cartes de visite*, and mail-order prints turned the immediacy of death into a marketable commodity and the exposed, anonymous corpse into a pervasive cultural presence."[4] In October 1862, a collection of battlefield photographs was exhibited at a New York City gallery owned by Matthew Brady. "The Dead of Antietam" was the work of several of Brady's assistants, including Whitman's friend Alexander Gardner. The poet probably attended the exhibit; he was still residing in New York prior to his departure for Fredericksburg in December.[5] His brother, George, fought at Antietam, where more than 23,000 men were killed, wounded, or missing in action.[6]

Photographers' access to Civil War casualties was unprecedented. Antietam was the first battlefield in American history to be photographed before the dead were buried.[7] The images captured were both traumatically provocative and exceedingly popular. The *New York Times* (October 20, 1862) described these posthumous photographs as the equivalent of repatriating the dead: "Mr. Brady has done something to bring home to us the terrible reality . . . of war. If he has not brought bodies and laid them in our dooryards and along the streets, he has done something very like it."[8] Crowds circulated among the images, captivated by the melancholy spectacle: "hushed, reverent groups stand around these

6. Timothy H. O'Sullivan, *Incidents of the War. A Harvest of Death*, 1863. From Alexander Gardner's *Photographic Sketchbook of the War*. Library of Congress. Prints and Photographs Division.

weird copies of carnage, bending down to look in the pale faces of the dead."[9] The reporter observed the images' capacity to evoke reverential fear, a "terrible fascination [that] draws one near these pictures, and makes him loath to leave them."[10] This morbid thrall recalls Whitman's descriptions of the hospital's "magnetic sights," and his inability to leave behind the bodies of the wounded at Fredericksburg.[11] Like the viewers in Brady's gallery, Whitman was "chained by a strange spell that dwells in dead men's eyes."[12]

Following the success of Brady's exhibition, the demand for war photographs soared. Alexander Gardner became the nation's first embedded war photographer, camping alongside Union troops for long periods of the conflict. Gardner's *Photographic Sketchbook of the War* appeared in 1866, a collection of eerily pastoral memorial images.[13] (See figure 6.) As Mark S. Schantz has illustrated, Gardner's photographs "repackaged the destruction of war" in scenes that conjured familiar associations with antebellum postmortem photography.[14] He rearranged corpses in poses that revised their violent deaths to correspond with the dominant aesthetic convention of posthumous portraiture: peaceful slumber.[15] Although mutilated and dismembered bodies were prevalent on battlefields, they rarely appear in photographs.[16] While images of corpses scattered across the killing fields served as a testament to the horrors of war, they also enacted a memorial function, allowing viewers the lost opportunity of the posthumous gaze.

In the pages of the *Atlantic Monthly,* Oliver Wendell Holmes narrated the tendency of certain survivors to resist the photographs' invitation to voyeurism, seeking instead to enact psychological containment. Like Whitman, Holmes's connection to Antietam was personal—he had recently traveled to the battlefield in search of his son. Having actually seen the war's aftermath, he had no desire to reencounter its casualties on the walls of a gallery.[17] Holmes struggled to "bury" the traumatic flashbacks provoked by viewing Brady's photographs:

> Let him who wishes to know what war is look at this series of illustrations. . . . It was so nearly like visiting the battlefield to look over these views, that all emotions excited by the actual sight of the stained and sordid scene, strewed with rags and wrecks, came back to us, and we buried them in the recesses of our cabinet as we would have buried the mutilated remains of the dead they too vividly represented.[18]

Holmes's desire to repress these violent images within the mind's "cabinet" recalls the human specimens enclosed within display cases at the Army Medical Museum. Like the specters that arise from Whitman's notebooks, conjuring the "actual sights" of battles and surgeries, Holmes describes the capacity of photographs to emotionally resurrect "the actual sight of the stained and sordid scene." This phantasmal afterimage must then be "buried" within the "cabinet" of the psyche, just as the subject wishes it were possible to inter the "mutilated remains" that the photograph "too vividly represented." Holmes uncannily foreshadows Roland Barthes's theory that the camera reframes the present vision as a dead past.[19] By virtue of its permanence in outlasting the object it represents, these photographs convey upon the corpse a perpetually unburied status. The survivor is thus incapable of entirely incorporating the dead, as long as the image remains. A surrogate crypt must therefore be produced. For Holmes, this grave site is the psychic "cabinet" of repression; for Whitman, it is an enduring afterlife within the pages of his book.

The Dark Bequest

Reading Whitman's war elegies alongside the emerging technologies of photography and embalming, this chapter examines the preservation compulsion that swept postbellum America, culminating with the assassination of Abraham Lincoln. Whitman articulates a pervasive anguish for the unburied dead and those who suffered partial interments in mass or shallow graves, linking the psychological processes of incorporative mourning with the depth of the deceased's absorption into the earth. In "The Bravest Soldier," Whitman mourns a specimen who, "standing . . . for hundreds, thousands," epitomizes an idealized battle death:

> Likely the typic one of them (standing, no doubt, for hundreds, thousands) crawls aside to some bush-clump, or ferny tuft, on receiving his death-shot—there sheltering a little while, soaking roots, grass and soil, with red blood—the battle advances, retreats, flits from the scene, sweeps by—and there, haply with pain and suffering (yet less, far less, than is supposed) the last lethargy winds like a serpent round him—the eyes glaze in death—none recks—Perhaps the burial-squads, in truce, a week afterwards, search not the secluded spot—and there, at last, the Bravest Soldier crumbles in mother earth, unburied and unknown.[20]

Like Civil War-era photographers, Whitman allows his "Bravest Soldier" the battlefield anomaly of a good death. There is a distinctly pastoral quality about the scene: the soldier retreats to a "secluded spot" —a bower-like "bush-clump, or ferny tuft"—which "shelter[s]" him from the advancing battle, but also obscures his body from the "burial-squads." Shrouded in "lethargy," he experiences "far less pain and suffering . . . than is supposed." Whitman's melancholy is reflected not in the soldier's death, which is relatively peaceful despite the circumstances, but in the landscape that obscures his corpse. The soldier's blood drenches "roots, grass and soil"; his unburied body "crumbles in mother earth." This is not the slow absorption of burial, but the rapid decay of unincorporated death. Once "the beautiful uncut hair of graves," the grass is now "soaked" with blood. Nature's detoxifying and regenerative properties are depleted by this permeation. (See figure 7.)

Describing the Battle of Chancellorsville, Whitman portrays an uncanny discord between the impassive natural world and the carnage of war:

> The night was very pleasant, at times the moon shining out full and clear, all Nature so calm in itself, the early summer grass so rich, and foliage of the trees—yet there the battle raging, and many good fellows

7. John Reekie, *Unburied Dead on the Battlefield of Gaines' Mill*, 1862. Library of Congress. Prints and Photographs Division.

> lying helpless . . . the red life blood oozing out from heads or trunks or limbs upon that green and dew-cool grass. The woods take fire, and many of the wounded, unable to move . . . are consumed—quite large spaces are swept over, burning the dead also—some of the men have their hair and beards singed—some splotches of burns on their faces and hands. . . still many are not only scorch'd—too many, unable to move, are burn'd to death.[21]

Nature's serenity in the wake of the battle's massacre creates a startling ecological dissonance. The "early summer grass" is drenched with the "life blood" that seeps from dismembered bodies. A sense of fatal indifference underlies Nature's inability to alter her "pleasant" "calm" to suit this violent milieu. As fire engulfs the wounded and the dead, an apathetic moon oversees their burning bodies:

Then the camp of the wounded—O heavens what scene is this?—is this indeed *humanity*—these butcher's shambles? . . . There they lie, in the largest, in an open space in the woods, from 500 or 600 poor fellows—the groans and screams—the odor of blood mixed with the fresh scent of the night, the grass, the trees—that Slaughter-house. . . . One man is shot by a shell, both in the arm and leg—both are amputated—there lie the rejected members. . . . Such is the camp of the wounded—such a fragment, a reflection afar off of the bloody scene—while over all the clear, large moon comes out at times softly, quietly shining. . . . Amid the woods, that scene of flitting souls . . . the impalpable perfume of the woods—and yet the pungent, stifling smoke—shed with the radiance of the moon, the round maternal queen, looking from heaven at intervals so placid—the sky so heavenly . . . a few placid stars beyond, coming out and then disappearing—the melancholy, draperied night above, around.[22]

Formerly botanical mechanisms for cleansing and absorbing the dead, "the grass [and] the trees" have now become the war's abattoir—a repository for the "camp of the wounded" and the "rejected members" of battlefield amputations. Whitman paints a "bloody scene" of sensory disparity where "the impalpable perfume of the woods" mingles with the scent of burning flesh. The "placid moon" and stars illuminate this unfolding massacre. Most startling of all is the sense that the detoxifying potency of the grass has been "saturated" by the blood of soldiers. Rather than incorporating and atomically transforming their corpses, the forest has become a "slaughter-house," passively observed by the "melancholy" nightscape.

As Whitman's "Bravest Soldier" evokes, thousands of men perished in obscurity—thrown into trenches en masse, stripped of identifying possessions and then abandoned on the field, or hastily buried in unmarked graves. More than 40 percent of Union soldiers and far more Confederates were, in Whitman's words, united in death "by the significant word *Unknown*."[23] The war ruptured Whitman's ecoerotic connection with the anonymous dead, creating an anxiety around his inability to access their "specimen" bodies:

The dead in this war—there they lie, strewing the fields and woods and valleys and battle-fields of the South . . . the varieties of the *strayed* dead (the estimate of the War department is 25,000 national soldiers kill'd in battle and never buried at all, 5,000 drown'd—15,000 inhumed by strangers, or on the march in haste, in hitherto unfound localities—2,000 graves cover'd by sand and mud by Mississippi freshets, 3,000 carried away by caving-in of banks, &c) . . . the numberless battles, camps, hospitals everywhere—the crop reap'd by the mighty reapers, typhoid, dysentery, inflammations—and blackest of all and loathesomest of all, the dead and living burial pits . . . Some where they crawl'd to die, alone, in bushes, low gulleys, or on the sides of hills—(there, in secluded spots, their skeletons, bleach'd bones, tufts of hair, buttons, fragments of clothing, are occasionally found, yet) . . . the clusters of camp graves, in Georgia, the Carolinas, and in Tennessee—the single graves left in the woods or by the road-side (hundreds, thousands, obliterated)—the corpses floated down the rivers . . . some lie at the bottom of the sea—the general Million, and the special cemeteries in almost all the States—the Infinite Dead—(the entire land is saturated, perfumed with their impalpable ashes' exhalation in Nature's chemistry distill'd and shall be so forever, in every future grain of wheat and ear of corn, and every flower that grows, and every breath we draw)—not only Northern dead leavening Southern soil—thousands, aye tens of thousands, of Southerners, crumble to-day in Northern earth.[24]

This long, chaotic sentence catalogs the human and ecological devastation inflicted upon the country by the ravages of war.[25] Whitman's statistical account of the "varieties of the *strayed* dead" includes unburied and unfound bodies alongside corpses that escaped shallow graves and "floated down the rivers," coming to rest "at the bottom of the sea." His halting prose mirrors the scattered remains it commemorates, while his attempts to count and categorize casualties reveal a collector's instinct.[26] As "A Voice from Death" (1889) articulates, Civil War history was shaped by the efforts of Americans to number, bury, and in some cases rebury their dead:[27]

The gather'd thousands in their funeral mounds and
 thousands never found or gather'd.
Then after burying, mourning the dead,
(Faithful to them, found or unfound, forgetting not, bearing the
 past, here new musing).[28]

The poet implores a "resign'd, submissive" America to entomb the war's "cataclysm" "deep" within the nation's reunified heart.[29] Whitman's "infinite dead" are finally "distill'd" within "Nature's chemistry"; blossoms once again emerge from their mouths: "out of death, and out of ooze and slime, / The blossoms rapidly blooming, sympathy, help, love."[30] Yet, at pivotal moments throughout the postbellum canon, Whitman interrogates the capacity of the earth (and the psyche) to absorb so many "*strayed* dead."

Whitman's recurrent metaphor of blood saturating the ground is more than poetic symbolism. His descriptions are remarkably similar to field accounts of battle, influenced by the soldiers he encountered in the hospitals. At a loss for words to evoke the carnage they witnessed, soldiers relied on metaphors of sensory exhaustion.[31] For example, David Thompson, a private with the 9th New York who fought at Antietam, described the psychological effects of the battle as a hallucinatory visual experience: "The whole landscape for an instant turned slightly red."[32] Survivors recalled their inability to cross the battlefield without walking over the dead, their feet rarely touching the ground.[33] Others described the "discoloration" of the dead as decay took hold, echoing the racial hierarchies that underpinned the war: "The faces of the dead," wrote a Union Gettysburg soldier, "turned black—not a purplish discoloration, such as I had imagined in reading of the 'blackened corpses' so often mentioned in descriptions of battlegrounds, but a deep *bluish* black, giving a corpse with black hair the appearance of a negro."[34] Fear of decay and its effects was not limited to soldiers. Army surgeons described the "effluvia of putrefying corpses" and "poisonous fungi in the atmosphere" emanating from mass graves, which they blamed for the scourge of epidemic diseases.[35] Alongside photographers, another group of entrepreneurs sought to capitalize on the collective desire to view soldiers' remains. Embalmers proposed a solution to the war's posthumous diaspora, advertising a new technology that facilitated the repatriation and preservation of fallen soldiers.

Postbellum editions of *Leaves of Grass* form a "reminiscent memorial"—a nostalgic crypt in which to house the war's "untold histories" and "unfound" bodies.[36] Whitman no longer seeks to ecologically absorb and regenerate the dead, but to infinitely preserve them through textual "embalming" practices:

> Perfume therefore my chant, O love, immortal love,
> Give me to bathe the memories of all dead soldiers,
> Shroud them, embalm them, cover them all over with tender pride.
>
> Perfume all—make all wholesome,
> Make these ashes to nourish and blossom,
> O love, solve all, fructify all with the last chemistry.[37]

In "Ashes of Soldiers," Whitman asks the poem to do what was formerly nature's work—to "fructify" the dead with the "last chemistry" that erases decay, transforming their ashes into "wholesome" compost. Whitman's desire to "perfume" the dead echoes the era's medical fears of miasmatic contagion, analyzed in chapter 1, a hygienic risk that embalmers claimed to eradicate. The "chant" enacts the hospitality of funereal rites (shrouding and embalming the corpse), while the poet acts as medium, inviting his ghosts to remain eternally by his side.

Whitman returned to the themes of anonymity and embalming in "A Twilight Song" (1891). Upon his textual "deathbed," as the last edition came to be known, Whitman remained haunted by the "vacant names" of unknown soldiers:[38]

> As I sit in twilight late alone by the flickering oak-flame,
> Musing on long-pass'd war-scenes—of the countless buried
> unknown soldiers,
> Of the vacant names, as unindented air's and sea's—the unreturn'd,
> The brief truce after battle, with grim burial squads, and the
> deep-filled trenches
> Of gathered dead from all America, North, South, East, West,
> whence they came up . . .[39]

At the close of his life, Whitman focused on the "gathered dead," now symbolically interred within *Leaves of Grass*. While those memorialized remain anonymous, this poem obscures the unburied bodies that fill the pages of Whitman's war notebooks and *Memoranda*. Yet, this elegy for the "unreturn'd," quietly alludes to corpses that somehow evaded the "burial-squads"—whether abandoned on the battlefield, expelled from "deep-filled trenches" by erosion, or estranged from mourners in unmarked graves. Because these diasporic bodies remain unabsorbed, they are vicariously untouchable: the poet cannot access them through nature. A "chant" of invocation is needed to "recall" and "embalm" the war's "dark bequest":

> Your million unwrit names all, all—you dark bequest from all
> the war,
> A special verse for you—a flash of duty long neglected—your
> mystic roll strangely gather'd here,
> Each name recall'd by me from out the darkness and death's
> ashes,

> Henceforth to be, deep, deep within my heart recording, for
>   many a future year,
> Your mystic roll entire of unknown names, or North or South,
> Embalm'd with love in this twilight song.[40]

Whitman questions the limits of mourning when burial is absent, when graves overflow, and the ground runs red with blood. His chant emerges from a fear of *chaotic dispersal*: that the unknown dead would be lost to him forever, unincorporated through either ecological or psychological processes. To assuage the fear that his beloved "specimens" would be rendered inaccessible, he searched for textual mechanisms of interment. Whitman seeks not only to bury, but also to "embalm" his soldiers, to guard against the decay he once celebrated.[41] Even as soldiers' "unfound" remains escaped integration, *Leaves of Grass* incessantly shifts in search of some way to locate and collect them.

In addition to the exhibition of the cadaver as a specimen of pathological significance, the preservation of the body became increasingly important to survivors during the Civil War. Antebellum mourners perceived embalming as an unnecessary defilement of the corpse. Before the war, its primary purpose was to preserve cadavers for medical dissection. The war rapidly altered this perception, creating an urgent desire to prolong the visibility of the fallen. Embalming appealed to hygienic as well as sentimental concerns, offering a practical method for transporting bodies across long distances. Northerners who had means went to great lengths to retrieve the remains of their loved ones, often traveling to battlefields to reclaim their kin. Embalmers and undertakers could also be enlisted to repatriate the bodies of lost soldiers.[42]

Unsettling the sentimental façade of death as peaceful slumber, embalming entailed highly invasive postmortem procedures—evidence of rapid changes in cultural relations with the dead. Decay was usually arrested by arterial injection with chemicals including arsenic, zinc chloride, bichloride of mercury, and a cocktail of salts, alkalines, and acids. If these materials were unavailable, an alternative method involved hollowing the body's trunk and stuffing it with sawdust, powdered charcoal, or lime.[43] A Washington correspondent for *Frank and Leslie's Illustrated* (May 1862) published a graphic description of the procedure:

> The body is placed on an inclined platform, the mouth, ears, nose &c. are stopped with cotton; if wounded, cotton is placed in the wound, and a plaster is put on; an incision is made in the wrist, the attachment is made

8. Embalming surgeon at work on a soldier's body, date and location unknown. Library of Congress. Prints and Photographs Division.

> from an air pump, and fluid is injected into the arteries. The wound is then sewed up and the body is hoisted up to dry. . . . The operators say in four months the body will become solidified like marble, but no chance has yet been had to prove it.[44]

Embalming's intrusive technology eclipses the regenerative decay prevalent in antebellum editions of *Leaves of Grass*. After "Nature's chemistry" was saturated with the blood of soldiers, embalmers' human chemistry enabled physical intimacy with the dead, posthumously filling and suturing wounds. (See figure 8.)

Over the course of the war, Washington emerged as the "nation's embalming capital," due to the large number of hospitals located in the city.[45] A full-page advertisement taken out by Dr. F. A. Hutton, who operated at 451 Pennsylvania Avenue, illustrates the emphasis placed on preserving the body as a simulacrum of the living entity:

> Bodies Embalmed by us NEVER TURN BLACK! But retain their natural color and appearance; indeed, the method having the power of preserving bodies, with all their parts, both internal and external, WITHOUT ANY MUTILATION OR EXTRACTION and so as to admit of contemplation of the person Embalmed, with the countenance of one asleep. . . . Surgeons and all interested are cordially invited to call and examine specimens after Embalmed. . . . N.B. Particular attention paid to obtaining bodies of those who have fallen on the Battle Field.[46]

Hutton's advertisement demonstrates the lengths practitioners undertook to assuage concerns for the sanctity of the body and detract attention from invasive surgical details. Hutton appeals to the mourners' desire to imagine death as slumber and to engage in the ritually significant final viewing. Whitman's near-constant presence in the hospitals makes it unlikely that he would have been so naive regarding the procedural intricacies.

Ironically, given Hutton's emphasis on the "attention paid to obtaining bodies of those who have fallen on the Battle Field," embalmers regularly advertised their trade by displaying preserved bodies, often those of soldiers stolen from battlefields.[47] In the tradition of medical body snatching, Thomas Holmes, who had embalming premises in downtown Washington, Georgetown, and Alexandria, routinely absconded with unidentified corpses, which he exhibited for advertising purposes.[48] Whitman would almost certainly have seen such displays during his daily walks through the capital. In addition to the theft of corpses for capitalistic gain, embalmers also exploited antebellum fears of disease emanating from the bodies of the dead. An announcement in the Richmond press promises "persons at a distance" that the bodies of their loved ones would be "Disinterred, Disinfected, and SENT HOME" from "any place within the Confederacy."[49] This Civil War incarnation of the exhumation compulsion paints embalmers as medical resurrectionists who "disinter and disinfect" the dead, enabling their posthumous return. Embalmers' appeals to bereaved families echo Whitman's fears of miasmic contagion and the diaspora of the "*strayed* dead." Perceiving the detoxifying capacity of nature to be compromised by war's rampant slaughter, the poet searched for another mechanism to cleanse and preserve the war's anonymous decedents.

Double Graves

At the close of the Civil War, the entire nation seemed cloaked in the black draperies of mourning; death had visited almost every household. In the months following the armistice, Union officials turned their attention toward the war's unburied and unidentified casualties. Whitman's anxiety surrounding the scattering of soldiers' remains was shared throughout the country. The preservation compulsion that spawned the embrace of embalming technologies also prompted reburial initiatives that sought to reverse the anonymity of battlefield death. The cessation of combat allowed military officials to finally begin organizing the retrieval and reburial of those in mass or unmarked graves.[50] A Union quartermaster general's report submitted not long after Appomattox listed only 101,736 registered burials, approximately one-third of the Union dead.[51] To combat this posthumous diaspora, the postwar reburial program endeavored to locate, identify, and exhume Union dead throughout the South, resulting in the creation of a network of graveyards that would eventually become the national cemetery system.[52] In the end, the program successfully identified 58 percent of Union dead.[53]

In June 1865 Captain James Moore was sent to Wilderness and Spotsylvania "for the purposes of superintending the remains of Union soldiers yet unburied and marking their burial places for future identification."[54] Moore uncovered hundreds of skeletal remains subjected to the elements for more than two years. "By exposure to the weather," he wrote, "all traces of their identity were entirely obliterated."[55] Although Moore was prevented from relocating all of the bodies due to oppressive heat and "the unpleasant odor from decayed matter," he ensured that all were properly interred.[56] The captain estimated that he had arranged for the burial of approximately 1,500 soldiers (785 of which were identified), although the dispersal of bones made an accurate account impossible. Soldiers from the U.S. Colored Troops, not yet released from service, were primarily responsible for this gruesome labor.[57]

On December 26, 1865, the chief quartermaster of the Military Division of Tennessee, Edmund B. Whitman (no known relation to Walt), was reassigned from his current post and ordered to "locate the scattered graves of Union Soldiers" across the South.[58] The quartermaster approached this mission from the conviction that "a *knowledge and a record of every grave*" must be "*in the possession of some living person*."[59] He searched for surviving witnesses who possessed information regarding the location and identity of the dead. He began by distributing a circular soliciting information from "Surgeons, Chaplains, Agents of Sanitary and Christian Commissions, Quartermasters, Officers or Soldiers," which he published in 300 newspapers.[60] The document announced the government's intention to create a "record . . . of all Union soldiers who have been buried in the Rebel States," and solicited assistance in locating their graves.[61] Responses flooded in from soldiers throughout the country. Many furnished maps indicating the exact spot where a friend had been buried. Edmund Whitman received descriptions "so minute and accurate in the details, that any person could proceed with unerring certainty to the very grave."[62] Like Whitman, the quartermaster was repulsed by the environmental consequences of unchecked decay. The "entire country over which war has extended its ravages," he wrote, "composes one vast charnel house of the dead."[63]

In spite of their scale and grandiosity, memorialization projects failed to assuage the national melancholia. Whitman's writings on "the Million Dead" reveal traces of violent trauma permanently inscribed upon the landscape:

> In some of the cemeteries nearly *all* the dead are unknown. At Salisbury, N. C., for instance, the known are only 85, while the unknown are 12,027, and 11,700 of these are buried in trenches. A national monument has been put up here, by order of Congress, to mark the spot—but what visible, material monument can ever fittingly commemorate that spot?[64]

While Whitman speaks to the limits of a "visible, material monument," this designation leaves open the possibility of another type of memorial: internal, immaterial, and unfixed. *Leaves of Grass* seeks to enact the latter function. By virtue of its textuality, it is a material artifact. Yet, Whitman's insistence on the shifting liminality of the text (through unending alterations) speaks to the crucial portability and open-endedness of his project. His elegies for the "scattered" dead foreshadow Caruth's understanding of the transient permanence of trauma: "Trauma is not *locatable* in the simple violent or original event in an individual's past, but rather in the way that its very unassimilated nature—the way it was precisely *not known* in the first instance—returns to haunt the survivor later on."[65] Whitman's interrogation of the memorial's commemorative failure speaks to the impossibility of isolating trauma to a particular location. It is precisely this unassimilable placelessness that returns to haunt, attesting to the impossibility of traumatic return. Unlocatability affects both the phantom and the haunted subject: trauma has no fixed place within the past (it returns at any moment), just as the unfound dead have no known grave. In this sense, both the survivor and the ghost are members of a psychic diaspora, unable to return to the site of their traumatic origins.

In *Drum-Taps,* Whitman elegizes soldiers' private burial rituals, similar to those uncovered by Edmund Whitman in his search for lost graves. In "Vigil Strange I Kept on the Field One Night" (1865), he imagines a soldier returning to the battlefield under the cover of darkness to bury a fallen comrade:

> Vigil final for you, brave boy (I could not save you,
> swift was your death,
> I faithfully loved you and cared for you living—I think
> we shall surely meet again);
> Till at latest lingering of the night, indeed just as the
> dawn appear'd,
> My comrade I wrapt in his blanket, envelop'd well his
> form,
> Folded the blanket well, tucking it carefully over head,
> and carefully under feet;
> And there and then, and bathed by the rising sun, my
> son in his grave, in his rude-dug grave I deposited;
> Ending my vigil strange with that—vigil of night and
> battle-field dim;
> Vigil for boy of responding kisses (never again on earth
> responding);
> Vigil for comrade swiftly slain—vigil I never forget,
> how as day brighten'd
> I rose from the chill ground, and folded my soldier well
> in his blanket,
> And buried him where he fell.[66]

The scene is similar to thousands of covert burials that occurred throughout the war. As Martin Murray observes, Whitman may have based the poem on an incident he recorded in an earlier notebook: "William Giggee, Sept 18th, '62. I hard of poor Bill's death . . . shot on Pope's retreat—Arthur took him in his arms, and he died. . . . Arthur buried him himself—he dug his grave."[67] Civil War soldiers were terrified of dying anonymously, and battlefields bore the scars of survivors' hurried attempts to bury the fallen. Soldiers preparing for the Battle of the Wilderness on May 6, 1864, were faced with gruesome reminders that the Battle of Chancellorsville had been fought on the same ground a year earlier. "The men who had fallen in that fierce fight had apparently been buried where they fell, and buried hastily," wrote Frank Wilkeson, a Union artilleryman. "Many polished skulls lay on the ground. Leg bones, arm bones, and ribs could be found. . . . Toes of shoes . . . bits of faded, weather-worn uniforms, and occasionally a grinning, bony, fleshless face."[68]

While lacking the sentimental accouterments of domestic deathbed rituals, Whitman's battlefield vigil nevertheless offers a tender homage to the departed soldier. Particular attention is paid to wrapping the corpse in a makeshift shroud. An army blanket "envelop'd well his form"; the mourner "folded the blanket well, tucking it very carefully over head and carefully under feet," before burying him "where he fell," in a "rude-dug grave." The meticulous enclosure of the body within the blanket and the book enacts a preservationist function, like a botanical specimen carefully pressed between sheets of paper. Despite the "rude[ness]" of the grave, the textual shroud enfolds the remains. The "vigil strange" offers another possibility for the unfound dead, an alternative to abandonment or mass interment by strangers—burial by a "loving comrade."[69]

In "As Toilsome I Wander'd Virginia's Woods" (1865) the Whitman persona uncovers an anonymous grave protectively shrouded by fallen leaves:

As toilsome I wander'd Virginia's woods,
To the music of rustling leaves, kicked by my feet (for 'twas
autumn)
I mark'd at the foot of a tree the grave of a soldier;
Mortally wounded he, and buried on the retreat (easily all
could I understand);
The halt of a mid-day hour, when up! no time to lose
—yet this sign left,
On a tablet scrawl'd and nail'd on the tree by the grave,
Bold, cautious, true, and my loving comrade.

Long, long I muse, then on my way go wandering;
Many a changeful season to follow, and many a scene of life,
Yet at times through changeful season and scene, abrupt,
alone or in the crowded street,
Comes before me the unknown soldier's grave—comes the
inscription rude in Virginia's woods,
Bold, cautious, true, and my loving comrade.[70]

The grave is isolated in pastoral seclusion, obscured by autumn leaves reminiscent of the blanket of "tomb-blossoms" analyzed in chapter 1. The privacy afforded this grave site continues the motif of secret interment in the aftermath of battle. The epitaph inscribed on a nearby tree returns to haunt the poet on "crowded" city streets, yet this flashback is comforting rather than traumatic. It is a vision conjured to counter the knowledge of absent burials.

Further illuminating the erotic attachments underlying textual burial, both poems bear similarities to Whitman's correspondence with Thomas P. Sawyer, a soldier he nursed at Armory Square Hospital. Sawyer occupied the same ward as Lewy Brown following an injury at the Second Battle of Bull Run (August 28–30, 1862). By the time Sawyer was discharged after six months of convalescence, Whitman was enthralled with him. The poet drafted a series of letters throughout the following months, becoming increasingly distraught when Sawyer neglected to reply.[71]

> Dear comrade you must not forget me, for I never shall you. My love you have for in life ~~and~~ or death forever. I don't know how you feel about it, but it is the wish of my heart to have your ~~tru~~ friendship, and also if you should come ~~all right~~ safe out of this war, we should come together again in some place where we could make our living, and be true comrades and never be separated while life lasts——and take Lew Brown too, and never separate from him. Or if things are not to be—if you get these lines, my dear darling comrade, and any thing should go wrong, so that we ~~cannot~~ do not meet again, here on earth, ~~I wish you to~~ it seems to me (the way I feel now) that ~~I could n~~ my soul could never be entirely happy, even in the world ~~after this~~ to come, without you, dear comrade. ~~What I am writing have written is pretty strong talk, I suppose, but it is I mean exactly what every word I say.~~ * God bless you Tom, and preserve you ~~Dear brother~~ through the ~~dangers~~ perils of the fight.
>
> Good bye my darling comrade, my dear darling brother, for so I will call you, and wish you to call me the same.
> * And if it is God's will, I hope we shall yet ~~live~~ meet, as I say, if you ~~could~~ feel as I do about it—and if is destined that we shall not, you have my love none the less, whatever should keep you from me, ~~or however long our though it be for~~ no matter how many years.[72]

Whitman's painstaking revisions lend the letter a halting quality, as though the author were struggling to find the words to describe what he felt for, and desired from, Thomas Sawyer. One month later, the poet expressed anxiety concerning the tone of his previous letter, fearing that it may have appeared "strange and unusual": "My dearest comrade, I cannot, though I attempt it, put in a letter the feelings of my heart—I suppose my letters ~~look~~ sound strange and unusual to you . . . but I am only expressing the truth in them . . . . I do not expect you to return for me the same degree of love I have for you."[73] Whitman returned to the metaphor of unspeakability to portray his struggle to elucidate "in a letter" the depth of his "feelings" for his "Dear comrade." As many critics have noted, there is an underlying anxiety in Whitman's assurances that he does not expect reciprocity from Sawyer, "hop[ing]" only for "at least a little of the feeling I have about you. If it is only a quarter as much I shall be satisfied."[74]

Sawyer's reply to Whitman's initial letter (April 26, 1863) was formal and distant in comparison: "I fully reciprocate your friendship as expressed in your letter and it will afford me great pleasure to meet you after the war will have terminated or sooner if circumstances will permit."[75] Although Whitman sent several more letters throughout 1863, Sawyer did not respond again until January 21, 1864. His letter was brief, comprising only two paragraphs and including a rather vague apology for his sporadic correspondence: "Dear Brother, I hardly know what to say to you in this letter for it is my first one to you . . . I should have written to you before but I am not a great hand at written and I have ben very buisy [*sic*] fixing my tent for this winter and I hope you will forgive me."[76] The poet was distressed by the soldier's long silence, stating plainly in July 1863, "I can't understand why you have ceased to correspond with me."[77] In an earlier draft letter (April 26, 1863), Whitman chastised Sawyer for failing to collect articles of clothing from him before departing:

> I was sorry you did not come up to my room to get the shirt and other things you promised to accept from me ~~at the time~~ and take when you went away. I got them all ready . . . a good strong blue shirt and a pair of drawers and socks, and it would have been a satisfaction to me if you had accepted them. I should have ~~felt pleased at the~~ often thought ~~you~~ now Tom may be wearing around ~~your~~ his body something from *me* ~~down there in camp~~, and that it might contribute to your comfort, down there in camp on picket, or sleeping in your tent.[78]

The melancholic urgency of the poet's longing is evident in his desire to visualize Tom "wearing around his body something from *me*." This yearning to encircle the beloved's body recalls the blanket that shrouds the dead soldier in "Vigil

Strange." Overall, the letters reveal striking correlations with Whitman's poetic burials. His "wish" to live with Tom as "true comrades" recalls the unknown soldier's epitaph in the Virginia woods: "Bold, cautious, true, and my loving comrade." Whitman's desire "never [to] separate" from Sawyer and Brown, and his prayer for "God" to "preserve [Tom] through the perils of the fight" emphasizes the intractability of his attachment, and foreshadows his poetic appropriation of embalming as a preservationist technology. Whitman's protestations that his "soul could never be entirely happy" "if we do not meet again" recalls the mourner's elegy for his fallen comrade in "Vigil Strange": "I faithfully loved you and cared for you living—I think we shall surely meet again."[79] The line also hints at Whitman's "faithful" care for Sawyer while he was recovering at Armory Square.

Whitman grieved over the loss of the masculine democracy he envisioned in "Calamus" (1860) and glimpsed within war camps and hospitals, a phenomenon that began to dissipate as soldiers returned home, which for many also signaled a return to heteronormative relationships. While his poetic burials enact mourning rituals for the anonymous dead, they also serve another psychological purpose, allowing the poet to lay to rest the linguistic echoes of the living soldiers who eluded him. Embedding remnants of these love letters with his poems, Whitman buries a lost living comrade alongside a dead one, constructing a "double grave" that is both textual and actual:

> For the son is brought with the father
> (In the foremost ranks of the fierce assault they fell;
> Two veterans, son and father, dropt together,
> And the double grave awaits them.)
> . . .
> O strong dead-march, you please me!
> O moon immense, with your silvery face you soothe me!
> O my soldiers twain! O my veterans, passing to burial!
> What I have I also give you.[80]

Whitman often characterized his relationships with soldiers in familial terms: he invited Sawyer to call him "darling brother," and spoke of Erastus Haskell as his "poor dear son."[81] An uncanny kinship persists in the duality of "soldiers twain"—a father and son "passing to burial." Whitman constructs a "dirge for two," a "double grave" in which to contain absent comrades, living and dead. His practice of relocating poems between clusters recalls the exhumation and relocation of Union decedents to national cemeteries. In this sense the locale of the grave site, like the wandering ghost, is driven by diasporic upheaval. Whitman's revised scattering of poems also replicates the phytological process of decomposition and regeneration, disseminating the war dead throughout *Leaves of Grass*.[82] Yet, this textual burial does not come without a cost. As Adam Bradford argues, "Whitman constructed *Drum-Taps* to act as a repository for the bodies of the lost soldiers of the Civil War, a repository which, by virtue of its status as a textual commodity, could circulate throughout the populace and carry those bodies—previously lost and left on the battlefield—home to be reclaimed by a loving readership." Poetic incorporation paradoxically "resulted in Whitman stripping Civil War soldiers . . . of the identities that marked them as unique, irreplaceable human beings."[83] Bereft of their identifying markers and "spiritual character[istics]," the specimen as poetic specter loses something of the "animal vitality" so provocatively rendered in *Memoranda* and the hospital notebooks.[84] Transformed into a portal for the infinite losses of others, the specimen loses its specificity.

In the war's environmental aftermath, Whitman sought to restore an ecological framework of fertility and regeneration. "A Carol of Harvest, for 1867" returns to the theme of battle as a violent reaping prevalent in Gardner's *Photographic Sketchbook of the Civil War*.[85] The poet longs to turn away from the "ghastly ranks" and "unnatural shows of war," and to reclaim the dead as compost for "Fecund America" through this "song of the grass and fields."[86] While soldiers' corpses are assimilated into the earth's "infinite teeming womb," a furtive anxiety lingers; the poet remains unsettled by the brutal origins of the harvest's fecundity:[87]

> On the far-stretching beauteous landscape, the roads and lanes,
> the high-piled farm-wagons, and the fruits and barns,
> Shall the dead intrude?
>
> Yet the dead mar not—they also fit well in Nature;
> They fit very well in the landscape, under the trees and grass,
> And along the edge of the sky, in the horizon's far margin.[88]

The question of ghostly intrusion betrays the poet's melancholic unease. Although the dead "fit very well in the landscape," their incorporation feels somehow forced: "the dead mar not." This simple, declarative statement is far removed from the intricate synecdoche of the prewar editions. Whitman cannot forget, however much he might desire to, that the "trees and grass" into which the dead seem to "fit" so well were only recently drenched in their blood. It is impossible to ignore the fact that the dead *do* intrude, over and over, throughout the postwar editions of *Leaves of Grass*. Corpses are never far from the surface. Haunted by the anonymity and placelessness of so many fallen soldiers, Whitman returns to questions of unburial, diaspora, and haunting: is the ghost rendered homeless by the absence of a grave? Unable to find lasting resolution to his anxiety for the unknown and unburied, Whitman incorporates these "unfound" ghosts within the text as surrogate tomb. Following the war, his poetic gaze engages in a sweeping visualization of the landscape in an effort to rescue the bodies that escaped absorption.

The poet's adoration of the wounded and dying represents, according to Robert Leigh Davis, a physical and textual alliance with partiality. "Both poet and text remain partial, in-valid. What Whitman attempts, however, is to revalue that partiality—that 'partial recovery.'"[89] The poet demonstrates a recurrent insistence, throughout the editions, on the erotic relevance of absent bodies, including, finally, the posthumous specter of the author. The poet continued to search for mechanisms through which to inter the unfound dead while simultaneously maintaining access to them. Psychological internalization alone was impermanent and therefore inadequate—the dead would die with him. Whitman required an incorporative vehicle that would allow him to reconstruct the regenerative natural "chemistry" shattered by the traumas of war. *Leaves of Grass*, reconfigured as a textual "memorial," became that object.[90] Still, a unifying force was needed, a legacy through which to unite the scattered ghosts of war. The resolution to this crisis of diasporic haunting arrived as a panoramic ghost capable of representing a multiplicity of corpses. Similar to the poetic presence who merged fluidly with living bodies in the first edition, Whitman conjures a phantom large enough to contain the "Infinite Dead."[91] Initially this inclusive specter took on the messianic form of Abraham Lincoln, and, finally, the "apparition" of the poet himself.[92]

Dissected Leaves

The publication of *Drum-Taps* was halted by Lincoln's assassination. Whitman signed a stereotyping contract on April 1, 1865, intending for the volume to be distributed one month later.[93] The events that followed changed everything. The Confederate surrender (April 9) and the assassination of the president five days later (April 14) rendered *Drum-Taps* glaringly unfinished. Whitman added an additional poem in Lincoln's memory, "Hush'd Be the Camps To-Day," but it fell short of the coherence demanded by the tragedy. He decided to withhold *Drum-Taps*, although some copies escaped, both by accident and intention.[94] Lincoln's death required a far more elaborate memorial, an elegy reflecting his significance to the Union and his unprecedented national funeral. *Drum-Taps* and *Sequel to Drum-Taps* were published in October 1865. Bound into extant copies of *Drum-Taps*, the *Sequel* contained eighteen new poems, including "When Lilacs Last in the Door-Yard Bloom'd." This suture of the sequel into the original volume symbolizes the fusion of Lincoln's death with Whitman's memorialization of the war at large. As Folsom and Price observe: "In joining *Drum-Taps* and *Sequel*, Whitman created a book whose physical form echoed the challenges the postwar nation was facing as it entered the stormy period of Reconstruction. Whitman, too, was entering a period of poetic reconstruction, searching for ways to absorb the personal and national trauma of the Civil War into *Leaves of Grass*."[95] The first incorporation of *Drum-Taps* into *Leaves of Grass* (1867) was similar in structure; it was not initially absorbed but annexed, relatively intact, at the end of the volume. As Moon explains, full integration required additional time and psychic work that could not be done in the immediate aftermath of the war: "*Drum-Taps* and its sequel could not simply be incorporated into the next edition . . . as a new 'cluster' because the national history, the massive experience of state violence . . . out of which it is written remained at least for a time (1867–1871) literally and figuratively unassimilable."[96] In contrast to the prosthesis of this initial attachment, many *Drum-Taps* poems would later be scattered throughout the final arrangement of *Leaves*.[97] Similarly, *Memoranda during the War* (1876) was absorbed into *Specimen Days* (1882), as Whitman sought to integrate the war experience into the overarching narrative of his life.

Given Lincoln's status as Whitman's most revered democratic specimen, it would seem logical that his death should be the most difficult to absorb textually. Yet, despite Whitman's deep sadness at Lincoln's passing, his elegies for the slain president do not reveal the same melancholic quality associated with his writings on the legions of unknown soldiers. Unlike the slaughter of so many faceless men, Lincoln's public mourning symbolized, for Whitman, the surviving coherence of the Union:

> The tragic splendor of his death, purging illuminating all, throws round his form his head, an aureole that will remain and will grow brighter through time, while History lives, and love of Country lasts. By many has *this Union* been conserv'd and help'd; but if one name, one man, must be pick'd out, he, most of all, is the Conservator of it, to the future. He was

> assassinated—but the Union is not assassinated . . . The soldier drops, sinks like a wave—but the ranks of the ocean eternally press on. Death does its work, obliterates a hundred, a thousand—President, general, captain, private—but the Nation is immortal.[98]

Whitman positions Lincoln as a martyred savior whose death conserved the nation. The poet inters anonymous soldiers alongside the president, in the hope that the "tragic splendor" of his death will "purg[e]" lingering anxieties of unburial and "illuminate" the nation's renewed "immortality."

Americans responded to Lincoln's assassination in profoundly personal ways, reflecting the inherent domesticity of mourning rituals, even on the national stage.[99] Reverend Henry Ward Beecher explained the internalization of this national crisis: "Men were bereaved and walked for days as if a corpse lay in their house."[100] Given antebellum America's fascination with all things macabre, it is hardly surprising that Lincoln's body became a site of voyeuristic pilgrimage. Images of Lincoln's deathbed became a national collage of projected desires for the bodies of countless Union soldiers. Families who were unable to observe deathbed vigils vicariously consumed prints and engravings depicting the dying president surrounded by mourners.

Four days after the assassination, the White House opened its doors to thousands of spectators, who filed solemnly past Lincoln's open casket in the East Room. In addition to the official state funeral, churches across the Northeast held memorial services. Clement M. Butler, a Philadelphia minister, remarked on the unparalleled national spectacle: "It is a new thing, this *actual participation* of a whole nation in the funeral obsequies of its fallen chief."[101] After the White House service, Lincoln's body was transported to the Capitol building, where he lay in state beneath the recently completed iron dome.

Like many of the soldiers who served under his command, Lincoln's body eventually returned home, undertaking a panoramic funeral journey that both mirrored and eclipsed all those that preceded it. On April 21, Lincoln's casket, accompanied by the body of his young son William (who had died of typhoid fever three years earlier), was transferred to an ornate funeral car for the long journey to Springfield, Illinois. The train covered 1,662 miles over twelve days, retracing the inaugural route the president had traveled just over four years earlier, stopping at Northern cities along the way. Citizens gathered near railway stations and alongside the tracks to witness the train's passage.

The public retained unprecedented access to the president's corpse. In each of the designated cities, thousands attended funeral ceremonies culminating with the display of Lincoln's body. The procession ended on May 4, with an elaborate funeral at Oak Ridge Cemetery. Even then, Lincoln's burial was not final. The president's remains were interred in the cemetery's public receiving vault until a more elaborate tomb could be constructed. In December 1865, his body was relocated to a temporary vault built solely for that purpose by the Lincoln Monument Association, while the group considered designs for the permanent structure. The temporary vault was later demolished so that it could never house another corpse. Finally, on September 19, 1871, Lincoln's body was transferred to the partially completed crypt.[102]

As Americans sought to reconcile the war's catastrophic aftermath, they embraced rituals that imposed order over the chaotic slaughter. By the 1880s, the fetishization of fallen soldiers that spawned the National Cemetery Movement had evolved into a reminiscence industry. Impromptu rituals enacted by families, fellow soldiers, and strangers morphed into formal traditions of patriotic memorialization, evident in the proliferation of monuments and the tradition of adorning graves on Decoration Day, eventually known as Memorial Day.[103] The emergent nostalgia industry sought to capitalize on public fascination with the glorious dead, especially the martyred figure of the fallen president. Ironically, at a time when postmortem photography was increasingly influential, only one posthumous photograph of Lincoln is known to exist. Secretary of War Edwin Stanton ordered all photographs destroyed. One surviving image of Lincoln lying in state was discovered among Stanton's papers after his death. To keep up with the flourishing demand for images of the slain president, many counterfeit deathbed photographs were produced.[104]

Whitman's enthrallment with Lincoln's image was established long before his assassination. Most days, on his way to Armory Square Hospital, Whitman loitered on the corner of L Street or wandered slowly down Fourteenth Street at twilight, in hopes of catching a glimpse of Lincoln's open barouche.[105] He recorded sightings of the president throughout his residency in Washington: "I see the President almost every day, as I happen to live where he passes to or from his lodgings out of town." Whitman was familiar with Lincoln's routine—so familiar, in fact, that he claimed, "we have gotten so that we always exchange bows, and very cordial ones."[106] More than anything else in his writings on the slain president, it was Lincoln's face that captivated Whitman.

> I see plainly ABRAHAM LINCOLN'S dark brown face, with the deep cut lines, the eyes, &c., always to me with a deep latent sadness in the expression. . . . I saw the President in the face fully . . . and his look, though

> abstracted, happen'd to be directed steadily in my eye. He bow'd and smiled, but far beneath his smile I noticed well the expression I have alluded to. None of the artists or pictures have caught the deep, though subtle and indirect expression of this man's face. There is something else there.[107]

Like the poet's convalescent soldiers, there is something "unworld[ly]" in Whitman's portrayal of Lincoln: "the deep latent sadness" behind his melancholy eyes.[108] Most uncanny is the inability of portraits or photographs to depict a true simulacrum. Whitman's Lincoln is already spectral, unable to be captured by the eye or lens (save Whitman's own gaze). The explanation for this obscurity is also beyond articulation; Whitman states, simply, "there is something else there." A notation to this passage offers a detailed description of the first sighting, prophetically funereal in tone: "I shall not easily forget the first time I saw Abraham Lincoln. It must have been about the 18th or 19th of February, 1861."[109] The president appears clothed in mourner's black, ungainly and "uncouth" despite his regal bearing:

> The figure, the look, the gait, are distinctly impress'd upon me yet; the unusual and uncouth height, the dress of complete black, the stovepipe hat push'd back on the head, the dark-brown complexion, the seam'd and wrinkled yet canny-looking face, the black, bushy head of hair, the disproportionately long neck, and the hands held behind as he stood observing the people. All was comparative and ominous silence. The new comer look'd with curiosity upon that immense sea of faces, and the sea of faces return'd the look with similar curiosity.[110]

Once again, Lincoln's "wrinkled yet canny-looking" face is the focus of reciprocal "curiosity" as he returns the onlookers' stares in "ominous silence." This "sea of faces" foreshadows the collective national gaze that fixed on the president's corpse throughout his funeral, as embalmers strove to lengthen the period between the final breath and the first sign of decay. Like Whitman's enigmatic war specimens, the president's oddness is "impossible to depict" in photography or portraiture. Yet, Whitman knowingly traces (and retraces) the "deep-cut" grooves of Lincoln's face, creating a literary life mask that long outlives its subject.[111]

Whitman's elegy for Lincoln begins in the "door-yard," the threshold between public and domestic spheres. The mourner "break[s]" a "sprig" from the "lilac-bush tall-growing with heart shaped leaves of rich green." One "flower" is selected to symbolize the loss; this blossom is transferred to the "coffin that slowly passes."[112] In *Memoranda,* Whitman described the early spring rains that caused lilacs to flourish that year, creating a sensory association between Lincoln's death and blooming flowers:[113]

> Early herbage, early flowers, were out. (I remember I was stopping at the time, the season being advanced, there were many lilacs in full bloom. By one of those caprices that enter and give tinge to events without at all being part of them, I find myself always reminded of the great tragedy of that day by the sight and odor of these blossoms. It never fails.)[114]

Like the olfactory hallucinations that resurrected the war hospitals in Whitman's mind, the scent of lilacs "never fails" to recall Lincoln's death. Whitman was not alone in connecting the president's passing with the abundance of seasonal flowers, which were widely associated with the funeral procession in newspapers and incorporated into memorial *cartes de visite*. Sprays of lilacs adorned the caskets of the president and his son at the Capitol.[115]

Despite this floral abundance, the poem radically departs from the ecoerotic symbiosis of the first edition of *Leaves of Grass*.[116] Although "every leaf is a miracle," the regeneration that follows death is now contained by the boundaries of ceremonial mourning.

> Over the breast of the spring, the land, amid cities,
> Amid lanes, and through old woods (where lately the
> violets peep'd from the ground, spotting the gray debris);
> Amid the grass in the fields each side of the lanes—passing
> the endless grass;
> Passing the yellow-spear'd wheat, every grain from its
> shroud in the dark-brown fields uprising;
> Passing the apple-tree blows of white and pink in the
> orchards;
> Carrying a corpse to where it shall rest in the grave,
> Night and day journeys a coffin.[117]

While Whitman hints at fertility emergent from decay in the "violets" that puncture the "gray debris" of the forest floor, and the "shroud[s]" that surround "every grain" of wheat, he does not linger upon images of decomposition. The lilacs' floral profusion is confined within domestic borders. Although the funeral train passes through "endless grass," the scenes of regenerative decay so provocatively rendered in earlier editions have vanished. Pastoral images are overshadowed by a "great cloud darkening the land," leaving "cities draped in black":[118]

> With the show of the States themselves, as of crape-veil'd
> women, standing,
> With processions long and winding, and the flambeaus of
> the night,
> With the countless torches lit—with the silent sea of faces,
> and the unbared heads,
> With the waiting depot, the arriving coffin, and the
> sombre faces,
> With the dirges through the night, with the thousand voices
> rising strong and solemn;
> With all the mournful voices of the dirges, pour'd around
> the coffin,
> The dim-lit churches and the shuddering organs—Where
> amid these you journey,
> With the tolling, tolling bells' perpetual clang;
> Here! coffin that slowly passes.
> I give you my sprig of lilac.[119]

The States are personified as "crape-veil'd women," whose "mournful voices" unite in a collective "dirge." The "sea of faces" gathered to mourn Lincoln's passing echoes the "curious" reciprocity that Whitman first recorded between the president and the crowd.[120] The poem's structure as a consolatory elegy mirrors the "long, winding" funeral train's procession; in this sense, consolation is achieved because burial was ritualistic, public, and prolonged.[121]

In addition to the "saturat[ion]" of "nature's chemistry" by so many unburied soldiers, a more practical explanation exists for the absence of decomposition within the poem: the president's body was never meant to decay. In preparation for the body's exhibition throughout the funeral train route, the presidential remains were chemically embalmed by the Washington firm of Brown and Alexander, who had previously embalmed the body of Willie Lincoln in February 1862.[122] After the corpse was drained of blood, a chemical solution was injected via the femoral artery, lending the body what observers described as a "statuesque, marblelike appearance."[123] Details of the embalming process were widely publicized, exemplified in the *New York World*'s graphic account:

> There is now no blood in the body, it was drained by the jugular vein and sacredly preserved and through a cutting on the inside of the thigh the empty blood-vessels were charged with a chemical preparation which soon hardened to the consistency of stone. . . . All this we see of Abraham Lincoln . . . is a mere shell, an effigy, a sculpture. He lies in sleep, but it is the sleep of marble.[124]

The "sacred preserv[ation]" of the presidential body, unmarred by decay, was a symbolic banishment of the deteriorating wounds that haunted the postwar Union. The human remains displayed at the Army Medical Museum paved the way for the publication of graphic details of Lincoln's autopsy and embalming, a phenomenon that would have been unthinkable even a decade earlier.[125] Blood relics saved by the surgeons who autopsied the presidential corpse eventually found their way to the Army Medical Museum, which was relocated to Ford's Theatre after the assassination. Alongside the surgical tools that performed the postmortem, today visitors can still see locks of the president's hair, fragments of his skull, and Surgeon Edward Curtis's sleeve cuffs, stained with Lincoln's blood. This enduring fascination with Lincoln's death pageant recalls Whitman's prediction that the president's assassination would "incise" a lasting scar upon American history: "When centuries hence . . . the leading historians . . . seek for some personage, some special event, incisive enough to mark with deepest cut . . . this turbulent Nineteenth century . . . those historians will seek in vain for any point to serve more thoroughly their purpose, than Abraham Lincoln's death."[126] Whitman's rendering of the assassination as a historical wound, "deepest cut," recalls published accounts of the president's autopsy and embalming. Benjamin Franklin Morris's *Memorial Record of the Nation's Tribute to Abraham Lincoln* (1865) details the surgeons' quest to extract the fatal bullet: "The wound was on the left side of the head, behind, in a line with and three inches from the ear. . . . In the track of the wound were found fragments of bone which had been driven forward by the ball, which was imbedded in the anterior lobe of the left hemisphere of the brain."[127] In Whitman's "Lilacs," Lincoln's corpse is conserved elegiacally, not only by the intervention of medical science, but also by the permanent inscription of his undying face within *Leaves of Grass.*

Despite concerted efforts, signs of deterioration crept into Lincoln's face: the hindrance of decomposition was ephemeral at best. Harry P. Cattell, the principal embalmer, accompanied the body all the way to Springfield. Working with local undertakers, he fought against time to ensure that the corpse remained palatable for public viewing. Yet, conflicting accounts surfaced as to his success. While many spoke of the "sweet," "peaceful," and "natural" expression of Lincoln's posthumous face, others contradicted this narrative as decomposition altered the president's features.[128] By April 24, when the cortège arrived in Manhattan, Lincoln's complexion had a grayish-brown hue. One New York reporter observed, "The color is leaden, almost brown; . . . the cheeks hollow and deep pitted."[129] By the time the funeral train reached Springfield, the limitations of embalming were readily apparent. The undertakers who opened the coffin to prepare the body for viewing were distressed by the increasing darkness of the face and its sunken features. The embalmer finally allowed the local undertaker to conceal the evidence of decomposition with the application of heavy layers of "rouge chalk and amber."[130]

As Gary Laderman has observed, the heightened visibility of Lincoln's remains was paramount in his resurrection as the archetypal martyred leader.[131] Such nationalist agendas depended on the large-scale repression of embalming failures. Yet, evidence remained, scattered throughout newspaper accounts and firsthand observations, of the embalmers' inability to truly prevent decomposition; in the end it could only mask the inevitable. Accounts of the body's subversive decay disrupted dominant narratives that presented the Union as wholly without flaw. Whitman responds with a synechdochic elegy that unifies the war's scattered casualties:

> (Nor for you, for one alone;
> Blossoms and branches green to coffins all I bring:
> For fresh as the morning—thus would I chant a song for
> you, O sane and sacred death.
>
> All over bouquets of roses,
> O death! I cover you over with roses and early lilies;
> But mostly and now the lilac that blooms the first,
> Copious, I break, I break the sprigs from the bushes:
> With loaded arms I come, pouring for you
> For you and the coffins all of you, O death).[132]

In contrast to the "uncut" "grass of graves" in the first edition, Whitman offers Lincoln "bouquets" of roses, lilies, and lilacs. Blossoms are *broken* "from the bushes" and ceremonially laid upon the coffin. Lincoln's "sane and sacred" body does not descend into the earth, "teeming" with unknown corpses.[133] Burial is no longer characterized by absorption into and dispersal throughout the ecoerotic landscape; it has become a domestic construct. The corpse is enclosed within a "burial-house":

> O what shall I hang on the chamber walls?
> And what shall the pictures be that I hang on the walls,
> To adorn the burial-house of him I love?[134]

Whitman's adornment of Lincoln's tomb is in keeping with rituals of remembrance that began informally before the war ended, and were nationalized in the years following. On Decoration Day, eventually known as Memorial Day, Americans left flowers at graves and monuments commemorating the war dead. By the early 1870s, when spring arrived and flowers were in full bloom, "one could not live in or near an American city or village, North or South, and remain unaware of the ritual of decorating graves of the Civil War dead."[135] Lincoln's "burial-house" includes not only bouquets, but also "pictures" to "hang on the walls," recalling the presidential memento mori that ornamented countless private homes.[136] The reminiscence industry boomed in the aftermath of Lincoln's death. The funeral train was accompanied by an escalating demand for mourning ephemera such as *cartes de visite,* stereographs, and mourning jewelry. A *Pennsylvania Daily Telegraph* reporter remarked on the commercialization of the president's death: "They are making fortunes out of it—by selling badges of mourning with Mr. Lincoln's Photograph."[137] Photographers who had previously captured battlefield carnage now turned their lenses toward images of the slain president. On May 22, 1865, Coleman Sellers reflected on the photographic frenzy that swept the nation:

> During the past month the whole labor of photographers has been in one direction—the collection and reproduction of portraits of Mr. Lincoln, and pictures of the localities and incidents connected first with the fearful tragedy itself, and then with the sublime spectacle of the funeral train. . . . Photography has furnished countless pictures of Mr. Lincoln, to be worn as badges and preserved as mementos of our late revered President.[138]

9. I. L. Rogers and John P. Soule, *Lincoln, Skeleton Leaves*, 1874. Private collection.

The popularity of Lincoln memorabilia occupied a parallel position to the public discourse surrounding his embalmed body. In the years following his assassination, the compulsion to memorialize the president fused with the sentimental art of creating "skeleton" or "phantom leaves" using botanical embalming techniques. The arrangement of hollow floral wreaths, often decorating photographs of deceased family members, was a popular pastime for middle-class women. Articles containing instructions on how to master the complex process frequently appeared in periodicals such as *Godey's Lady's Book.* Internal plant matter was removed by fermentation or chemical alteration, leaving behind only the sheer exterior of the petal. The transparent leaves were then dried and arranged into still lives, usually under a glass dome. Photographs of these spectral wreaths were often reproduced as stereographs. Phantom leaves encircling images of Lincoln were especially popular.

Irene L. Rogers was the era's most prolific creator of skeleton leaves memorials. A photographic legacy of her work survives, the result of a collaboration with the Boston photographer John P. Soule. (See figure 9.) Yet she remains a critically neglected and misidentified figure within postbellum visual and mourning cultures.[139] Prophetically, Rogers anticipated that her legacy might prove as ephemeral as her creations. Challenging the domestic sphere of feminized mourning rituals, she filed a patent application for her "Processes of Skeletonizing Leaves":

> In the process of preparing leaves, flowers, seed-vessels, &c., for skeletonizing, the separation of the vegetable and fibrous material of the leaves

> by first softening in muriatic acid and water and subjecting to heat. . . . The fiber is then bleached in the following manner: Make a solution of one-fourth pound chlorate of lime to about eight quarts of rainwater; place the fibrous matter loosely in a glass vessel, cover with the above, and expose to the light until the fiber becomes thoroughly bleached. . . . The fibrous matter, after being colored in the ordinary way, is made into bunches or used . . . for ornamental purposes.[140]

Rogers's instructions for "the separation of the vegetable and fibrous material of the leaves" through exposure to chemicals such as muriatic acid recall public descriptions of the embalming preparations carried out on Lincoln's corpse: "The scalp has been removed, the brains scooped out, the chest opened and the blood emptied."[141] This "Memorial" of "Dissected Leaves" references the president's highly publicized autopsy, offering another example in which Whitman's poetics intersected with contemporary botanical practices.[142] Embalming leaves and enshrining Lincoln, Rogers and Whitman represent sexually marginal undercurrents at work in postbellum America's preservation compulsion. Rogers's "phantom leaves" and Whitman's "lilacs" seek to construct *botanical memorials* surrounding the *face* of Abraham Lincoln. Although Rogers's stereographs were published after "Lilacs," both cultural objects participate in a shared memorialization project, in which "phantom" flowers wreath the "sacred remains" of a corpse that will never decompose. The dissection and preservation of organic material as historical artifact attests to the collective desire for sentimental permanence: the reclamation of ephemeral, dead, or discarded objects in an effort to subvert decay. This converges with a commodity fetishism that infinitely reproduces and distributes such objects, which were often personal or temporal in their original incarnations. The legacy of Irene Rogers's floral altars to Abraham Lincoln has long outlasted history's memory of their creator, obscuring even her name.

Lincoln's embalmers endeavored to present the grieving nation with the appearance of an unaltered image that remained entirely whole. By attempting to situate mourning processes within a bordered location, memorials implicitly designate other spaces as existing beyond the boundaries of bereavement. They are designed to obscure the existence of unfound bodies. Yet, as we have seen, this posthumous diaspora may not be permanent. Exhumation is just another word for resurrection, and the dead have many ways of returning home. Lincoln's elegy offered Whitman's readers a way of collecting (and collectively mourning) the "*strayed* dead." In this elegiac crypt, Whitman constructs a "burial-house" large enough to contain the war's diasporic ghosts. Through the poet's "unclosed" eyes, we see "long panoramas" of suffering:

> I saw battle-corpses, myriads of them,
> And the white skeletons of young men—I saw them;
> I saw the debris and debris of all dead soldiers;
> But I saw they were not as was thought;
> They themselves were fully at rest—they suffer'd not;
> The living remain'd and suffer'd—the mother suffer'd
> And the wife and the child, and the musing comrade suffer'd
> And the armies that remain'd suffer'd.[143]

As his spirit accompanies the hermit thrush in flight, the Whitman presence is no longer "bound" by "sight." His disembodied eyes take in a series of "visions," including uninterrupted views of the retreating armies, and the exposed remains of the unburied dead. The battle's aftermath returns "in noiseless dreams," suggesting that these "torn and bloody" images often recur as traumatic reverberations. The tattered "shreds of the flags" mirror the soldiers' weathered corpses. The "white skeletons of young men" echo the grass that "transpire[d] from the breasts of young men" in the first edition of *Leaves of Grass*, now more similar to the "Burial Party" captured in Gardner's *Photographic Sketchbook of the War.*[144] (See figure 1.) Yet, in defiance of these harrowing scenes, Whitman revises the good death to include violent assassination and battlefield slaughter; the poet insists that these scattered remains are "fully at rest," and "suffer'd not."

In spite of his elegiac efforts, the "myriads" of "battle-corpses" Whitman encrypts alongside Lincoln refused to stay buried; their "phantoms" are resurrected throughout the postwar editions of *Leaves of Grass*. In "Pensive and Faltering" (1868, 1871) the dead are represented as the "only living, only real," while the poet is rendered spectral:

> Pensive and faltering,
> The words *the dead* I write,
> For living are the Dead,
> (Haply the only living, only real,
> And I the apparition, I the specter).[145]

As Whitman's soldiers are entombed within shifting incarnations of *Leaves of Grass*, the text itself becomes an object of libidinal mourning. Following the war, communion with the dead is no longer engaged through ecoerotic decay but via textual afterlife: preservation replaces regeneration. The ever-changing book follows the wandering ghosts whose bodies eluded burial. Both author and reader, as war voyeurs, remain powerless to assuage the suffering that is its consequence. Whitman's postwar haunting entails a partial detachment from the ecoerotic landscape he once celebrated, a land now "teeming" with unmarked graves.[146] Textual incorporation inherits a memorial function akin to both the botanist's practice of "skeletonizing leaves" and the postwar reburial program. Whitman removes the site of haunting attachment from the grave and relocates it within the book—mourning has migrated beyond burial. The specimen has become the specter, whose habitat is both psychic and textual, portable and intrinsic.

5 Autopsy and Afterlife
Anatomical Celebrity

In the end, Whitman bequeathed his body not to the grass he loved, but first to the scalpel, and then to the tomb. Less than a year after the final edition of *Leaves of Grass* was published, the poet died at sunset on March 26, 1892. The cause of death was bronchial pneumonia, the final complication of tuberculosis contracted in the Civil War hospitals.[1] It took nearly thirty years, but Whitman at last became a casualty of war, alongside his beloved specimens. The poet would surely have enjoyed the following obituary from the *New York Mirror,* which cast him as a "magnificent specimen" whose "extraordinary vitality" was destroyed by his hospital service:

> While nursing wounded soldiers after the bloody fights of the Wilderness in 1864, Whitman contracted hospital malaria and was never himself again. The trouble which began here ultimately culminated in his paralysis. Whitman was forty-two years old when he went into camp and hospital to nurse soldiers. He was a magnificent specimen of manhood physically, a perfect picture of strength and natural grace. Since 1873 he has been simply a broken down man, waiting for the death which was all too long in coming. . . . Sweet death tarried long. Persons of extraordinary vitality live in spite of themselves under wasting disease, even when it is their dearest wish to depart. The strong, deep breathing body fights mechanically and yields only inch by inch. A weak physique would have succumbed years ago to the infirmities under which Walt Whitman suffered and lingered nearly twenty years, writing marvelous poems when his disease gave him respite enough.[2]

This passage echoes Whitman's own perception of the origins and progression of his illness, up to a point. "During the war I possess'd the perfection of physical health," he wrote in *Memoranda*.[3] As we have seen, in Whitman's deterioration began in the war hospitals, where he sacrificed "my health . . . my body—the vitality of my physical self" for the soldiers that he nursed:[4] "I suppose I should have been free of all of this today—free at least in part—if in those last years 63-4-5 I had gone off to a place of safety, avoided the hospitals . . . taken special care of my own person: but here I am, sick, nearly gone, and I do not regret what I did."[5] Not only did Whitman refuse to regret the "cost" of his hospital work, he considered the exchange of his health a fair trade for the "boys" who became integral to *Leaves of Grass*.[6]

Whitman's decline steepened in 1873, after a series of strokes left the poet partially paralyzed. Yet he was never a "broken" man, at least not psychologically. As Loving writes, "a study of the poet's temperament from 1873, when his health crisis began, and even earlier, during the war, when he faced occasional problems of serious proportion, suggests that he had a remarkable capacity to look beyond his medical situation."[7] Horace Traubel reports occasional periods of depression and frustration related to pain and immobility from 1888 onward, but these complaints quickly recede in almost every instance to engaging, even optimistic conversations that range far beyond the confines of the poet's "caged" bedroom on Mickle Street.[8] While he was "waiting for death" in the final months of acute suffering, this was often expressed with an absence of fear and an accompanying spirit of mirth and curiosity reminiscent of the haunting optimism of "Song of Myself": "to die is different from what any one supposed, and luckier."[9] In January 1891 Traubel wrote, "Walt very frankly expresses his anxiety to die, to shake off this burden, which increases and is heavier with each day."[10] From his own deathbed, the poet recalled witnessing soldiers' welcome deaths that arrived "like an invisible breeze after a long and sultry day," bringing not only relief from pain but even "a sort of ecstasy."[11]

During his twilight years, Whitman attracted a devoted cohort of disciples who were enthralled by his famous "personal magnetism" and dedicated to the promotion and protection of his literary reputation, both at home and abroad.[12] Whitman had always been receptive to the ardent younger men who gravitated toward him, from John Burroughs and William O'Connor in the Civil War years to Oscar Wilde, who twice journeyed to Camden in the early 1880s to pay homage to the poet.[13] One man in particular was ever-present during the Camden era. Horace Traubel was only thirty-three when the poet died, but he had known him for almost twenty years. They met in 1873, not long after Whitman moved into his brother George's Camden home. Beginning in March 1888, Traubel served as Whitman's constant companion and literary assistant, keeping meticulous records detailing the poet's life and letters, which he would eventually publish as *With Walt Whitman in Camden* (1906–96). It took nine volumes over ninety years to publish Traubel's narrative of Whitman's last four years.[14] This complex history is not limited to the volumes' lengthy publication. Traubel obscures the clarity of authorship and unsettles the nature of textuality. Is this a transcription, a biography, or a literary memoir? Did the student eclipse the master to publish "the most wayward, spontaneous, fragmentary book ever printed," as Whitman wrote of his own *Specimen Days*?[15] Lingering questions remain as to how accurately Traubel captured Whitman's voice, and whether the poet fully understood his intentions.[16] William Sloane Kennedy, another of Whitman's disciples, wrote that the poet would "probably have desired to have him privately shot if he had known what he was going to do after his death."[17] Traubel's public airing of the poet's private thoughts and correspondence led to long-standing controversies for Whitman's friends, readers, and scholars. For example, Traubel's account of Whitman's remarks about Silas Weir Mitchell exemplify the drastic consequences this posthumous publication had for those who found themselves unflatteringly represented. Yet, for all its complexities and caprices, Traubel's work is invaluable to Whitman studies, not only as a lens into the poet's final years, but also as a literary artifact that demonstrates the shifting subjectivities of memory and its afterlives.

This chapter establishes Traubel's account of Whitman's autopsy as a climactic moment in the collision of scientific and sentimental practices upon the body at the close of the nineteenth century. The disciple elegiacally captures the convergence of mourning and medicine, as the scalpel opens Whitman's body to yield "unexpected fruits."[18] Through his narrative of anatomical reverence, Traubel figures Whitman's autopsy *as* afterlife. His inventory of the removal of Whitman's organs bestows adoration on each part, independent of the whole, reminiscent of Whitman's concern for the severed limbs of soldiers. Following this anatomical "benediction," Traubel began to transfer his attachment from the body to the book, swiftly turning his attention toward Whitman's "literary effects."[19]

At the time of his death, Whitman's growing celebrity already included devotional societies across the United States, as well as Britain, Canada, and Australia, due in no small part to the tireless efforts of his disciples as well as the poet's lifelong flair for self-promotion.[20] His mounting American influence was reflected in the sheer volume of obituaries and public tributes upon his death. In keeping with the facets of the good death, many newspapers reported that Whitman died "peacefully": after his "long struggle with death, the end [came] like a child asleep."[21] The *New York World* described the poet's "face [as] majestic in its repose, [which] showed that he had met death in a calm and unfaltering spirit."[22] Another obituary memorialized the poet's departure as akin to the open-endedness of a book to which the reader returns, again and again: "The closing of his life was like the closing of a finished book, and those who cared for the man and his work are aware that in memory they can take up the book again and re-read it, from its brave, fearless, bold and untried beginning to its noble, eloquent, magnificent end."[23]

Whitman's deathbed vigil was elaborate and prolonged; in some sense it had been going on for months.[24] At the end, he was surrounded by vigilant caregivers: Dr. Alexander McAllister (his physician), disciples Traubel and Thomas Harned, Mary Oakes Davis (his housekeeper), and Warren "Warrie" Fritzinger (who assisted Davis with his care).[25] Warrie and Mary Davis washed the body while the undertaker was summoned.[26] Before the corpse was penetrated by the scalpel, Traubel ensured that the poet's face and hands were preserved in plaster. Thomas Eakins arrived on Sunday, March 27, and began work on Whitman's death mask. An ancestor of postmortem photography, the death mask aimed to capture the fleeting period between the final breath and the first sign of decay. By the nineteenth century, this ritual was associated with physiognomy, the pseudoscience aimed at discerning spiritual qualities by visibly analyzing the body, most often facial features and the shape of the head. In *Essays on Physiognomy* (1775), Johann Caspar Lavater encouraged the scientific study of the corpse, arguing that in the repose of death, features are more revelatory:

> The dead, and the impressions of the dead, taken in plaster, are not less worthy of observation [than the living]. Their settled features are much more prominent than in the living, and the sleeping. What life makes fugitive, death arrests; what was indefinable is defined. All is reduced to its proper level; each trait is in its true proportion.[27]

Eakins had painted Whitman's portrait in 1888; the poet admired the artist's unflinching depiction of his old age, likening the image to "a poor, old, blind, despised & dying king."[28] Famous for his realistic dissection scenes in *The Gross Clinic* and *The Agnew Clinic*, Eakins embodied the intersection of art and medicine. He attended anatomy classes at Jefferson Medical College in order to better depict the human form, and molded bronze casts from articulated cadavers.[29] At the Pennsylvania Academy of Fine Arts, Eakins insisted that all his students, including women, study dissection.[30]

The artist and his assistants worked for over three hours. Traubel remained present for much of this time, beginning the voyeuristic worship that would continue throughout Whitman's autopsy.

> Walt's face serene and sweet and composed. . . . How like one of the grand classic pictures of gods, with the hands calmly folded and that strange yellow-white, and peace everyhow lined however the eye looked! They worked and worked—I watched and watched. In from the north the gray light—outside the beating rain—the room, so long dedicated to his sacred work, still redolent of his nature. I could catch the faint odor of his hair. I touched his hand. Though cold it was yet somewhat pliable . . . W.'s head lay towards the window. The light played a strange beauty into his hair, and the pallor was no way painful . . . . W.'s serene face and folded hands and bared shoulders, as a god stretched out on god's own altar, dead.[31]

Echoing Whitman's public obituaries, Traubel sees evidence of Whitman's good death in the serenity and composure of his face. The room in which the cast is molded, the scene of Whitman's "sacred work," bears lingering traces of his presence. The atmosphere, "still redolent of his nature," flatters the poet, casting a "strange beauty" over the beloved corpse "stretched out on god's own altar." Traubel's Whitman is a departed deity, his body an abandoned "temple" to which the disciple must "keep close . . . until the final toll is paid."[32]

Traubel's devotion to the poet's postmortem features recalls Whitman's fascination with Lincoln; just as Whitman revered Lincoln's living face, Traubel worshiped Whitman's dying face. Likewise, Whitman's authorial apparition supplants Lincoln as the panoramic specter of the final edition of *Leaves of Grass.* Traubel's posthumous encounter references consolatory mourning narratives of death as divine slumber, and reveals the tenacity of his connection to the departed poet:

> W. lay stretched on a stretcher. I went into the room, uncovered and kissed him. The body was clean. The face already assumes a repose and majesty. . . . The body was already getting rigid. Eyes beautifully sweet and lips closed. Hand not nearly so fallen away as other members. He lay there in the light, his splendid head seen at its noblest and all the history of his tumultuous years wiped away by the touch of peace. The strange quiet smote me. I leaned over and kissed his forehead (oh! that kiss!

and the afternoon's kiss, the life just gone!). . . . Even the livid face was alluring—and it lay there like some grand old god pictured to the soul or memoried out of our loving and immortal friendship—no more—no more![33]

Traubel's description is in keeping with the physiognomic theory that the face, in the repose of death, revealed an innate truth: "his splendid head seen at its noblest." The posthumous "allure" of Whitman's "livid face" prompts Traubel's parting kiss, recalling Whitman's wartime accounts of receiving soldiers' "dying kiss[es]."[34] This fetishistic reverence for the body did not stop at the surface; the disciple's adoration pierced the skin to worship blood and bone as Traubel witnessed Whitman's dissection and dismemberment.

Unexpected Fruits

Later that same evening, Whitman's physician, Dr. Alexander McAllister, and colleagues Daniel Longakre and Henry Cattell arrived to perform the autopsy, which had been postponed due to George Whitman's vehement objections. According to reports in the *New York World,* George would only consent to the postmortem "if he were satisfied any scientific end was to be made . . . but he felt that the only purpose of such a course of action on the part of the doctors was the satisfaction of a professional curiosity."[35] The poet's brother left Mickle Street at noon following his final visitation with the body; after his departure, the doctors and executors deemed that the autopsy would go ahead after all. Traubel met with Longaker and Cattell earlier that afternoon in Philadelphia, reporting that he "found from them that W. had indeed consented in December . . . to a post-mortem after death. This may obviate George's objection." The doctors also told Traubel that Richard Maurice Bucke had assured them that "there would be no difficulties placed in the way" of an autopsy. Cattell then said, "Of course we could not do anything without the family's consent. And yet what a thing it means for science and knowledge," to which Traubel replied, "And for man." Yet there is no evidence that he contacted George Whitman again prior to the autopsy, which was rescheduled for 6:00 p.m. at Mickle Street.[36]

Significantly, George did not issue an unqualified objection. He expressed an aversion to his brother's dissection as a "professional curiosity," yet maintained respect for scientific discovery.[37] As we saw in chapter 1, antebellum Americans perceived dissection as a desecration of the corpse that had dire consequences for the soul in the afterlife. Yet autopsies were usually performed not for experimental purposes, but to discover the cause of death for persons deemed important enough to warrant such an investigation. As Kenneth Nystrom argues, nineteenth-century dissection can be read as a form of structural violence, whereas autopsy was not necessarily seen as such.[38] Whitman's autopsy was performed not only to ascertain his cause of death, but also for medically experimental purposes. Reminiscent of newspaper articles detailing Lincoln's autopsy, on March 28, 1892, the *New York World* featured a story provocatively titled "The Dead Poet's Brain: A Report That He Bequeathed It to the Anthropometric Society," detailing George's objections and the medical case in favor of a postmortem:

> [The doctors] believed an autopsy on Whitman would be of great value to science. He suffered from bronchial pneumonia and had survived for months, when his condition at one time indicated he could not survive five hours. He suffered the mysterious pain on his left side for years, which had been diagnosed as cancer. It was desirable to know whether this was true or not. It was also desirable to know whether there were any evidences in his brain of the strokes and paralysis from which he suffered years ago, and other facts in the interest of medical science might be shown.[39]

Brian Burrell and Sheldon Lee Gosline have uncovered archival evidence that illuminates "the strange fate of Whitman's brain," a century-old anatomical mystery.[40] Burrell makes a compelling case that Whitman was a member of the American Anthropometric Society. Founded in 1889 by prominent doctors interested in neurology, the Brain Society, as it was also known, aimed to collect and study the brains of eminent men donated by pre-mortem bequests.[41] The anatomist Joseph Leidy was among the founders. Whitman's doctors William Osler, Silas Weir Mitchell, and his son John were also members, and may have recruited the poet.[42] "In certain circles," Burrell writes, "leaving one's brain to science had become downright fashionable."[43] Whitman may have been seduced by the chance to become an anatomical celebrity. His brain was supposed to join those of other prominent intellectuals at the University of Pennsylvania's Wistar Institute. Yet this cerebral fame was not to be, at least not in the manner that the society promised.

On December 5, 1908, the Philadelphia *North American* ran a story titled "Brain Research by Philadelphia Anatomist Startles Science," in which Anthropometric Society member Edward Spitzka claimed that the quality of the mind could be measured by the shape of the brain. This theory failed to gain credibility, and Spitzka sank into medical obscurity, except for one detail included in the story: an offhand remark that Whitman's brain had slipped from a laboratory assistant's hands, sustained irreparable damage, and been discarded: "Unfortunately, not even the pieces were saved."[44] As Burrell establishes, this anecdote sparked controversy about the mistreatment of the bard's celebrated brain. Spitzka himself admitted that the incident resulted from "carelessness in handling."[45] Whitman's furious disciples immediately began inquiries to determine how the specimen came to such an end, and why no one was notified.[46] The *Camden Courier* reported Traubel's incensed response: "I cannot understand it. The fact that such an institution should permit the care of such a precious property to an attendant who probably had no idea of the value of what he was handling is bad enough. But that they should permit the brain to be lost and then fail to notify the executors of Walt Whitman's estate, I consider worse."[47]

In spite of the executors' outrage, the ambiguity surrounding who was responsible for the destruction (and disposal) of Whitman's brain remained unresolved until 2014, when Gosline acquired the diary of Henry Cattell, the Wistar Institute's head pathologist, who, according to Traubel, placed Whitman's brain into "his gupsack" after it was extracted.[48] The pathologist confessed in his diary to ruining Whitman's brain, not by dropping it, as Spitzka had suggested, but by accidentally leaving it in an unsealed specimen jar overnight.[49] Without the necessary conservation measures, Whitman's specimen was lost to science and history. Cattell remained haunted by the incident for the remainder of his life, though this was perhaps more from fear of professional fallout than guilt.[50]

In both Spitzka's published remarks and Cattell's private confession, "carelessness" is the common thread. The poet's anatomical specimen was not given anything like the care that he lavished on his hospital cases, or the preservationist fame bestowed on their medical museum counterparts. The doctors who fought to obtain Whitman's brain despite his brother's objections then treated it with casual negligence and discarded it as medical waste. No longer an object of "professional curiosity," the ruined brain was cast aside, reminiscent of the carelessness shown by antebellum anatomists toward the resurrection and disposal of dissected cadavers. The cause of the brain's destruction—exposure in an unsealed specimen jar—eerily recalls the unburied war dead whose bodies were likewise uncovered and subjected to the elements.

Yet before we judge Cattell too harshly, it must be acknowledged that wet specimens are fragile and laborious to maintain. As Gosline writes, "there are several ways to pickle a brain, involving a variety of injections and baths." Cattell's preferred method required between two and six months to complete—ample time for something to go wrong:

> An open jar, bucket, or wash-basin is one-quarter filled with absorbent cotton, and Muller's fluid is added until the vessel is about one-half filled. The brain, after being removed from the body and weighed, is carefully placed in the center of the vessel and more fluid is added, until the brain is well covered. If this be done there is no danger that the brain will decompose, even in summer. . . . The position of the brain is altered on the next day and the fluid changed. . . . The fluid is changed again on the third day, then every other day for three successive times, twice a week for the next three weeks, and once a week for the next three weeks.[51]

While the intricacies of this process reveal the high possibility for human error, Whitman's brain was not alone in suffering from scientific negligence. In a 1906 report to the society's members, Spitzka acknowledged that a number of "elite brains" had been severely damaged or lost entirely. The founder Joseph Leidy's brain had been incorrectly weighed. Andrew J. Parker's specimen was overexposed to a hardening agent and crumbled to pieces.[52] If Whitman's brain had survived, it would have been examined, sketched, perhaps molded into a plaster cast, and then forgotten in the basement of the Wistar Institute, alongside the other society specimens. As Burrell writes, "such has been the fate of almost all so-called 'elite' brains—oblivion."[53] Once reduced to anatomical material, the body—even the celebrity body—is no longer seen as entirely human. Recalling the words of Leidy's eulogist, "a corpse cannot say 'I am a man'"—not even when that corpse belonged to one of the era's most prolific and hyper-eloquent writers. In the end, the physical specimen extracted from the "poet of the body" proved far more fragile and ephemeral than his textual ghost.[54]

Whitman's autopsy was conducted between 6:10 and 10:00 p.m. on March 27, in the rear parlor of the poet's home. Traubel remained in the room throughout the procedure, cataloging the removal of each organ: "The heart stood alone in its perfection and strength. Everything else was impaired."[55] The poet's left lung had entirely collapsed, while the right retained only one-eighth of its breathing capacity.[56] Whitman's brain was "extracted, and seemed without hurt"; the doctors remarked on the "magnificent symmetry of the skull." Traubel observed, "The wonder of the doctors as operations proceeded seemed to grow. Once

Cattell said, 'This man must have lived weeks and weeks simply by force of will power.' I put in, 'And serenity.'"[57] In spite of the surgical violence inflicted upon the corpse that aroused in him so much devotion, Traubel witnessed the entire autopsy, remaining long after Warrie Fritzinger and Thomas Harned departed:

> To hear the claw and dip of the instruments—to see the skull broken and opened and the body given the ravening prey of the investigator had its horrors—then its compensations. I looked beyond and saw science, man, with benediction sweet. . . . Somehow I could not have gone home, leaving them at this work, or avoiding. I seemed to hear an injunction out of space, 'Keep then close to the temple till the final toll is paid.' And so I braved and threw that inner protest which so closely attended me throughout. To these men body and brain yielded unexpected fruits.[58]

Traubel silences his instinctive protests at the "horrors" of Whitman's posthumous wounds, commanded by a hallucinatory voice, "Keep then close to the temple till the final toll is paid." Traubel's inability to depart and "leaving them at this work" recalls Whitman's devotion to the hospitals' "magnetic yet terrible" sights.[59] The poet's cadaver is a "temple," housing sacred relics that multiply as each organ is removed. The disciple must keep watch throughout these anatomical excavations.

Traubel describes the autopsy with the same adoration that marked his last encounter with Whitman's intact corpse. Despite the dismemberment of the poet's divine body, "the ravening prey of the investigator" delivered certain "compensations." Body and brain yield "unexpected fruits" of scientific discovery. Traubel believed Whitman's brain would prove, through scientific study, to be extraordinary. Traubel suggests that Whitman's legacy would include an anatomical afterlife as a medical specimen, alongside his literary celebrity and spiritual divinity. "Look[ing] beyond" the "the claw and dip of the instruments," Traubel sees the autopsy as an anatomical "benediction," a final act of corporeal worship for the "poet of the body."[60] Rather than obliterating the body's opportunity for a divine resurrection in the afterlife, in Traubel's eyes the autopsy is an act of worship at the temple of the "exquisite corpse." Whitman's articulated cadaver merges with Traubel's melancholically incorporated ghost to become a fantastically mourned phantom that, in Maria Torok's words, the ego must "preserve carefully . . . in the unconscious": "the ego looks for this exquisite corpse continually in the hope of one day reviving it."[61] Traubel's efforts to preserve and revive Whitman's "exquisite corpse" entailed guarding and expanding his literary celebrity, and textually encrypting the poet within his own lifelong, "fragmentary" book.[62]

After the cadaver was opened to "investigators" and its treasures thoroughly examined and recorded, Traubel turned his gaze toward the conservation of Whitman's literary estate. On March 29 the poet's executors, Traubel, Bucke, and Harned, took possession of his papers (many of which resided chaotically on the bedroom floor). They packed the documents "pell-mell" into barrels which were then sent to Traubel's home to be sorted and archived.[63] Traubel was already extremely protective of Whitman's work, writing in the immediate aftermath of his death: "As to disposition of W.'s literary effects we urge caution, which is about all that can be done now."[64] Alongside the disciple's copious notes, these papers underpin *With Walt Whitman at Camden.* Traubel was by far the most prolific of the executors, publishing the first three volumes prior to his death in 1919, and leaving behind manuscripts for the remaining six.[65] In 1890 he founded *The Conservator,* a journal dedicated to Whitman's legacy, which he edited and published until his death. The *Conservator*'s motto, written by Traubel in 1893, echoes the disciple's depiction of the poet's body, as well as his body of work: "Moral conscientiousness, crystalline, without flaw, not godlike only, entirely human, awes and enchants forever."[66] Traubel published three books of Whitmanian poetry in his lifetime. All the while, he corresponded with Whitman devotees across the globe, uniting an international network of allies who worked to advance the poet's reputation.[67] Yet even the most ardent of his disciples could not surpass the poet himself as devoted curator of his "literary effects."

That Spectral Ring

Throughout 1891 Whitman devoted his remaining energies toward the production of a final volume of *Leaves of Grass*, in what he envisaged as its definitive form.[68] Having narrowly escaped death several times already, a sense of urgency compelled him to expedite the book's publication. Writing to his friend John Johnson in September, Whitman described his drive to "to finish out (bind) L of G with 'Good Bye' & last of all 'Backward Glance' & shall then let it go as completed as I can make it."[69] As the poet hurried to get the book to press, he recycled sheets from the 1888 printing, supplemented by annexes of "Sands at Seventy" (1888), "Good-Bye My Fancy" (1891), and "A Backward Glance O'er Travel'd Roads" (1888).[70] Initial copies were simply bound in plain wrappers, so that Whitman could personally send the book to close friends before he died.

Published by David McKay and bound in green cloth, the spines of the final volumes are gold-stamped "Leaves / of / Grass / Complete / 1892" above Whitman's signature, echoing the poet's insistence that his life's work was at last complete (or at least "as completed as I can make it").[71]

McKay had been Whitman's publisher since 1882, when he took over the business from Rees Welsh & Co., during which time he became Whitman's friend and ardent supporter. Nevertheless, as Folsom has established, "Whitman's very protective literary executors . . . did not fully trust McKay and were also anxious to control Whitman's fate and the future publishing of his books themselves."[72] Enhancing the links between his body and his book, Whitman told Harned that he did not expect his executors to remain loyal to McKay: "My spark'll go out any day now: I don't want to tie you fellows up: you may find reasons for going to another publisher. I wouldn't advise you to go but I wouldn't put my corpse in your way if you were disposed to make a change." After McKay's contract expired in 1895, the executors did indeed withhold future publication rights from him.[73]

In the "deathbed edition" Whitman sought to shape not only the past, but also the future—to secure and control his own posthumous celebrity. The cultural adhesiveness of Whitman's legacy speaks to the potential permanence of literary afterlives: the book outlasts its author, speaking continually in his absence. During the final year of his life, Whitman's fixation on the elaborate construction of his own tomb mirrors his annex to the final edition, "Good-Bye My Fancy." Originally published in the spring of 1891 as a pamphlet, Whitman conceived of "Good-Bye My Fancy" as an appendix to *November Boughs*.[74] Highlighting the book's purpose as a vehicle for psychological and textual incorporation, Folsom observes, "this notion of an 'appendix' added to an 'annex' stretches the figurative language of expansion and incorporation about as far as it can go."[75] Loving describes "Good-Bye My Fancy" as "Whitman packing his literary bags for eternity." This auto-elegiac annex is filled with poems reckoning with the author's imminent death. Yet, as Loving writes, "the tone is wonderment instead of fear."[76]

After working on the project throughout the fall of 1891, Whitman sent Bucke an advance copy of the final edition on December 6, with an accompanying note announcing: "L. of G. *at last complete*—after 33y'rs of hackling at it, all times and moods of my life, fair weather & foul, all parts of the land, and peace & war, young & old."[77] The volume was finalized just prior to the poet's final collapse. Less than a week later, he wrote again to Bucke, complaining of various ailments.[78] His health began declining rapidly in November, and by December 18 he demonstrated symptoms of bronchial pneumonia. Although doctors did not believe he would last a week, he lingered for a further three months, throughout which he remained deeply concerned with his poetic legacy. As his body deteriorated, Whitman wondered about the public reception of the final edition and fixated on Arthur Stedman's forthcoming anthology of his poems.[79] David S. Reynolds has described the poet's involvement with the arrangement of this anthology as "one more effort on Whitman's part to tailor his image to suit the ever elusive American public . . . just as he had been controlling and reshaping his public image from the start."[80] In his final months, Whitman was less concerned with his declining health than with the afterlife of his work. "Good-Bye My Fancy" is Whitman's attempt at achieving a good *textual* death: departing stoically and peacefully, surrounded by mourners, leaving behind a literary artifact that would survive its author.

In his "Preface" to "Good-Bye My Fancy," under the heading "Concluding L. of G.," Whitman justified the following collection of poems, arguing in favor of the literary relevance of "lingering-dying" words: "During the last two years I have sent out, in the lulls of illness and exhaustion, certain chirps—lingering-dying ones probably (undoubtedly)—which now I may as well gather and put in fair type."[81] As a collection of last words, "Good-Bye My Fancy" contributes to Whitman's literary self-fashioning; in the end, the poet turned from collecting war specimens and specters to collecting his own life's work. In his essay on Whitman, J. M. Coetzee elaborates on the selective function of the collected edition: "To bring out a Collected Poems does not mean to republish all the poems one has written in a lifetime. By convention, the collector is entitled to revise old poems and quietly omit those he or she no longer cares to acknowledge . . . to shape one's own past."[82] Whitman defied potential critics, arguing for the inherent worth of these "last droplets" that "stain" the "conclusion" of "a long dusty journey":[83]

> Had I not better withhold (in this old age paralysis of me) such little tags and fringe-dots (maybe specks, stains) as follow a long dusty journey, and witness it afterward? . . . In answer, or rather defiance, to that kind of well-put interrogation, here comes this little cluster, and conclusion of my preceding clusters.[84]

Whitman expresses trepidation not only regarding the critical reception, but also the literary merit, of his "old age collation." In conversation with Traubel on March 6, 1891, he remarked that the cluster was not to be classified as a "work" alongside *Leaves of Grass* as a whole: "It is hardly to be dignified as 'work': it is simply a last drop, a leave-taking, my farewell."[85] Nevertheless, Whitman deemed these "concluding drops" significant by virtue of their finality.[86]

The poet ends his last "Preface" not with meditation on his own mortality, but with recollections of his years spent "visiting and waiting on wounded and sick army volunteers . . . in hospitals or fields south of Washington City."[87] He closes by questioning the capacity of the Union to "realize what itself cost," and positioning his text as a "reminiscent memorial" that continually speaks to this unrealizable loss.[88] It is, Whitman argues, in the capacity of "last words" to reflect on all that preceded them, and to attest to the final, fleeting thoughts of their bearer, that their literary value lies.[89]

In order to conjure a textual presence that could outlast death, the poems of "Good-Bye My Fancy" interrogate the transience of reality. Reminiscent of the phantoms that emerged from Whitman's bloodstained war notebooks, "apparitions" haunt its pages. Yet it is not these specters, but the "solid things" of the living world that are "non-realities":

> A vague mist hanging 'round half the pages:
> (Sometimes how strange and clear to the soul,
> That all these solid things are indeed but apparitions,
>   concepts, non-realities).[90]

In presenting solidity as "vague" and insubstantial—as apparitional—Whitman simultaneously attests to the tangibility of the ghost. "The Pallid Wreath" reflects upon the status of the text as memorial object, and the poet's reluctance to relinquish sentimental ties to the departed: "Somehow I cannot let it go yet, funereal though it is. . . . One withered rose put years ago for thee, dear friend."[91] The "funereal" "wreath" recalls the "bouquets of roses," "lilies," and "lilacs" offered to Lincoln's coffin, though the flowers have now "faded":

> But I do not forget thee. Hast thou then faded?
> Is the odor exhaled? Are the colors, vitalities, dead?
> No, while memories subtly play—the past vivid as ever;
> For but last night I woke, and in that spectral ring saw thee,
> Thy smile, eyes, face, calm, silent, loving as ever:
> So let the wreath hang still awhile within my eye-reach,
> It is not yet dead to me, nor even pallid.[92]

Despite the passage of time and the effects of decay and "exhal[ation]" upon the wreath, it is not "dead"; the "memories" it evokes remain as "vivid as ever." The "past" is more provocative than the exhausted present. Once again, the sentimental and the "spectral" overshadow the actual: "it is not dead to me, nor even pallid." The wreath as "spectral ring" speaks to the cyclical continuity of the nostalgic object: even at its most dejected, its capacity for resurrection is not exhausted, but enhanced. It remains a lingering testament to the lasting power of both apparition and artifact.

An anonymous review in the *New York Tribune* (August 16, 1891) titled "Whitman's Farewell: A Melancholy Book" alludes to the text's "funereal" rhetoric of stoic departure: "A dreadful photograph resembling nothing so much as a death-mask serves as grim frontispiece to this ultimate publication by Walt Whitman. . . . There is a melancholy flavor about the whole of it, though the old man tries very hard to be cheerful."[93] Like the "death mask" photograph described by the reviewer, the text enacts a preservation function; it seeks to infinitely inscribe the authorial "farewell." The cluster is Whitman's anticipatory reckoning with his own imminent death, a vehicle for the self-fashioned haunting of its author. The title poem figures the poetic "self" as a specter that resists materiality and revels in finality, who clings to a "meaning" withheld from beyond the grave:

Good-bye* my fancy—(I had a word to say,
But 'tis not quite the time—The best of any man's word or say,
Is when its proper place arrives—and for its meaning,
I keep mine till the last).[94]

In the nineteenth century, the word "fancy" was "synonymous with imagination": "the process, and the faculty, of forming mental representations of things not present to the senses; chiefly applied to the so-called creative or productive imagination, which frames images of objects, events, or conditions that have not occurred in actual experience . . . (especially as denoting attributes manifested in poetical or literary composition)."[95] While this definition is overtly relevant to Whitman's rhetoric, another equally valid (if obsolete) usage exists: "a spectral apparition; an illusion of the senses."[96] Whitman's "fancy" can therefore be read as a literary specter that the poet *leaves behind*. The poem acts as Whitman's farewell not only to his audience, but also to his own "fancy." This literary ghost remains to haunt "future printing[s]" and their readers, even as the poet takes his leave. Whitman's notation to the word "Good-Bye" further elucidates the auto-elegiac significance of the conclusion, the "*last word*":

> Behind a Good-bye there lurks much of the salutation of another beginning—to me, Development, Continuity, Immortality, Transformation, are the chiefest life-meanings of Nature and Humanity. . . . Why do folks *dwell so fondly on the last words*, advice, appearance, of the departing? Those last words are not samples of the best, which involve vitality at its full, and balance, and perfect control and scope. But they are valuable beyond measure to confirm and endorse the varied train, facts, theories and faith of the whole preceding life.[97]

Whitman's "Good-bye" entails both finality and futurity; it speaks to the resonance of rhetorical departure, behind which "lurks much of the salutation of another beginning." This dual emphasis is enhanced when read alongside the following note, which appeared on the copyright page of the 1891–92 edition:

> As there are now several editions of *Leaves of Grass*, different texts and dates, I wish to say that I prefer and recommend this present one, complete, for future printing. . . . The subsequent adjusting interval which is so important to form'd and launch'd work, books especially, has passed; and waiting till fully after that, I have given (pages 423–438) my *concluding words*.[98]

Both notes attest to the haunting capacity of the printed word as both beginning and ending, enacting a Derridean form of cyclical return: "the return of the ghost as text, or the text as ghost."[99] Whitman's "concluding words" embody the echolaic power of finality: perhaps not the best but that which comes *last*. Speaking of the "unbearable" inheritance of textual legacy, Derrida insists that the geography of such a language is "unlocatable" in its inevitable silence: "Such testimonies survive us, incalculable in their number and meaning. . . . Already they survive us, keeping the last word—keeping silent."[100] By "keep[ing]" his meaning "till the last," Whitman ensures that he withholds the final word. In an attempt to control his own legacy, the poet offers this plea to listen to the last, as well as the best, of his words. He presents the "value" of these "concluding words" as confirmation of "the whole preceding," insisting that even when the best words are left behind, a spectral resonance exists in the language of departure.

Before Long It Will Justify Its Builder

As he arranged the publication of the final edition from his deathbed, the poet also designed his own "burial-house," which was completed around the same time.[101] Due to his growing celebrity, Whitman was offered burial plots in several cities. He chose Camden's Harleigh Cemetery because of its pastoral setting and proximity to his home on Mickle Street.[102] Harleigh was a product of the rural cemetery movement, a consequence of the grave robbing and sanitation anxieties that influenced Whitman's early poetry and prose. The cemetery was arranged in a series of landscaped gardens and winding paths. To maintain its rural ambience, no fences or enclosures were allowed; head and footmarkers larger than eight inches were also prohibited. However, large mausoleums, such as the one Whitman envisaged, were accepted.[103] The cemetery's director offered Whitman a free plot of his choosing in exchange for a poem. The poem remained unwritten, but Whitman claimed his plot nevertheless. In December 1889 he arrived at the cemetery in a borrowed carriage and cheerfully selected his grave site.[104] Although cemetery officials encouraged him to designate a prominent location, he journeyed into the woods and indicated a secluded spot on the hillside near Cooper's Creek for the construction of his mausoleum.[105]

On September 29, 1890, Whitman included a rough drawing of the tomb in a letter to Bucke.[106] The sketch depicts a simple domestic structure with a large door and a pitched roof, surrounded by design specifications and scenic descriptions under the heading "Walt Whitman's Burial Vault": "20 x 30 feet on a sloping wooded hill vault heavy undress'd . . . blocks grey granite—unornamental." (See figure 10.) Like his careful "decoration" of Lincoln's "burial-house," Whitman paid meticulous attention to the structural and interior materials of his own tomb.[107] Adjacent to the drawing, on the left-hand side, Whitman noted a description of the tomb's "surroundings": "trees, turf, sky, a hill, everything crude and natural." Whitman modeled his tomb on William Blake's engraving *Death's Door*, which he initially encountered in 1881 while reading Alexander Gilchrist's *Life of William Blake*.[108] The etching depicts an aged, bearded man entering the wide doorway of a stone enclosure. As this stooped figure crosses the threshold of "Death's Door," a resplendent youth sits atop the structure, encircled by rays of light from the rising sun, symbolizing the soul's immortality.[109] After selecting his plot, Whitman engaged a contractor to build the tomb based on his design.

Within the cemetery's idyllic setting, Whitman constructed a burial house that was a veritable fortress. As the *Camden Courier* reported, "nestled on the hillside, kept green by a gurgling brook, sheltered by an aged oak tree and its knarled branches, and protected by a massive structure of granite, the mortal remains of Camden's poet will rest until the sounding of the trump on the last day."[110] Whitman's mausoleum is an imposing stone structure comprising 72.5 tons of granite enclosing 600 square feet, recessed deep into the surrounding earth. The tomb is enclosed by an iron gate secured with a bronze lock.[111] Whitman had initially designed a granite door weighing 2,200 pounds, but it was too heavy to be hung.[112] Inside, the vault contains eight catacombs of white marble weighing 5.75 tons; the walls are lined with 15,500 bricks.[113] The initial building contract was $4,000, more than twice the price of Whitman's Mickle Street home.[114] This impenetrable tomb was designed to unite Whitman's family in death. The poet told his sister, Hannah: "It is my design to gather the remains of our dear father and mother and have them buried here in the tomb I built for myself."[115] Reminiscent of the postwar reburial movement, Whitman resurrected and reinterred his family within this granite tomb. His parents were moved from their graves in Brooklyn and Camden, and in time the mausoleum also housed the remains of siblings Hannah, Eddy, and George and his wife, Louisa, and their infant son.[116] As Kaplan writes, the poet also, "in a lasting assertion of self, merged their identities with his."[117] Walt Whitman's name, alone, was etched over the doorway.

Something other than vanity compelled Whitman to construct this monument alongside the unrelenting drive to "complete" *Leaves of Grass*.[118] Although Whitman chose a pastoral cemetery, he elected to be interred above ground, breaking his initial promise to leave his body to the "dirt."[119] Rather than allowing his body to be absorbed into nature, he chose enclosure within a granite sepulcher.[120] This stark grey mausoleum stands in stark contrast to Whitman's initial, auto-elegiac pact: "I bequeath myself to the dirt to grow from the grass I love, / If you want me again look for me under your bootsoles."[121] Whitman's self-entombment is a radical departure from the ecoerotic regeneration prevalent in the prewar editions. Far from facilitating the poet's incorporation into the earth, the stone tomb and its marble catacombs prevent, at all costs, the author's dissemination into the landscape. The tomb was designed to protect his body from "crumbl[ing] in the soil of mother earth."[122] By surrounding his corpse with tons of marble, bricks, and granite, Whitman inhibits the event that was once his ideal: incorporation into nature through corporeal decay.

Whitman witnessed the violation of many graves in his lifetime. He grew up near the derelict cemeteries of his ancestors in Long Island, "depress'd mounds, crumbled and broken stones, cover'd in moss."[123] In New York, he saw the graves

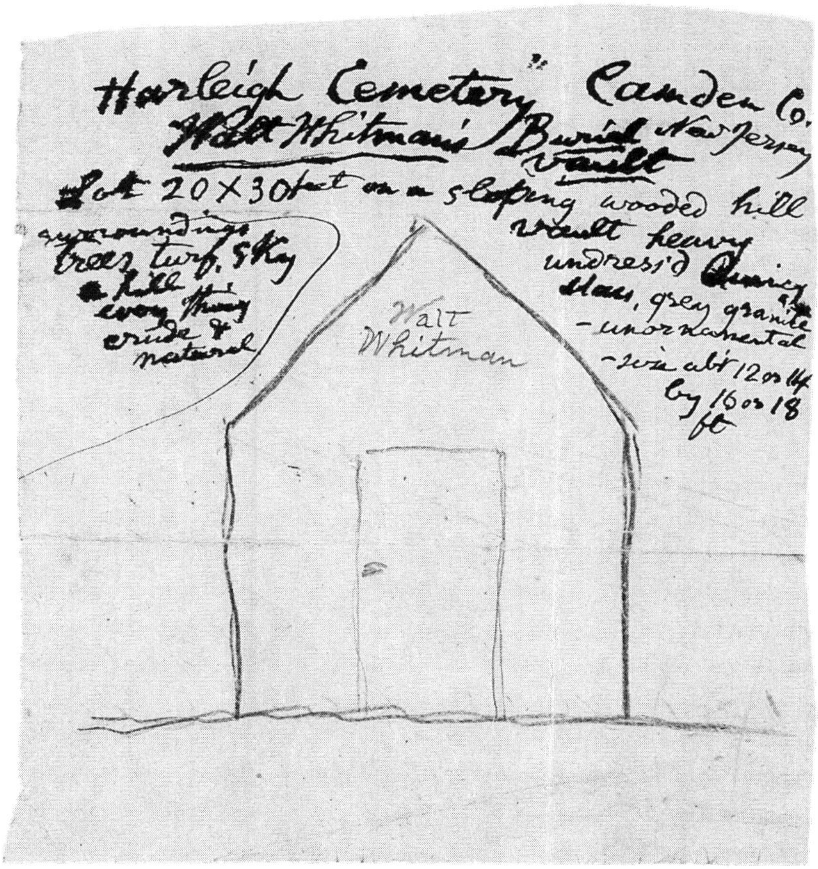

10. Walt Whitman, burial vault design, 1890. Library of Congress. Manuscript Division.

of Revolutionary soldiers disturbed by the city's relentless expansion.[124] During the antebellum years he reported with horror on the exploits of body snatchers. Throughout the Civil War, he witnessed the diaspora of "the strayed dead" whose unburied bodies littered battlefields and became lost to rivers.[125] Even Abraham Lincoln's consecrated tomb was not inviolate. In 1876 thieves broke into the president's mausoleum at Springfield; they were apprehended just as the casket was about to be opened.[126]

Whitman's formidable tomb was designed to guard the poet's remains from more than human disturbances. As chapter 4 demonstrated, the earth had been "saturated" by the blood of "countless" soldiers, compromising its regenerative consolation.[127] Could the earth, now "teeming with corpses," be trusted to "filter and fibre" the poet's blood into his descendants' veins?[128] Was there a more permanent method of preserving the poet and his work? Whitman wanted a secure resting place for his book and his body to await their legacies. Like Lincoln, Whitman would not go into the ground, but into the crypt. Both tomb and text arise from the same psychological impulse, a desire for memory. In Torok's words, through melancholic incorporation "the corpse is entombed in a fast and secure burial place, awaiting resurrection." Whitman's local newspaper, the *Camden Courier*, captured the crypt's resurrectionist tendencies in an article describing the construction of "the rugged granite tomb in which all that is mortal of the Rugged Old Poet will await the sounding of the trump."[129] Whitman's tomb is more than a response to the earth's "saturation" with the blood of soldiers; like the author's deathbed "completion" of *Leaves of Grass* and its auxiliary texts, the impenetrability of the vault is an act of legacy creation. The *Camden Courier* predicted that Whitman's mausoleum would "outlast the ordinary vaults of any monument ever constructed."[130] The poet and his disciples hoped that the same would hold true for *Leaves of Grass.*

Whitman's belief in the tomb's lasting significance is evident in a conversation recorded by Traubel on November 5, 1891: "W. says, 'It has a grim background. But before long it will justify its builder.' What did he mean? I wanted to hear more. He only said, 'It is a thing not to be disputed about, of course,' and left it."[131] Whitman's elusive testament to the canonical function of the tomb is reiterated by Traubel's longing for further elaboration. Whitman asserts the unwavering belief that both tomb and text will survive to "justify" their "builder." In spite of the author's celebrity status, *Leaves of Grass* had yet to achieve the cultural success Whitman envisaged for his life's work. Though his readership had expanded considerably in the postwar years, particularly in transatlantic and international contexts, *Leaves of Grass* was not yet installed alongside the family Bible in every American home.[132] Whitman's attention to the construction of his tomb and final text attests to his unwavering faith that America would inevitably awake to the overarching significance of *Leaves of Grass*, even if he did not live to see the day. Whitman's final response to Traubel's inquiry prevents further interrogation as to his meaning: "It is a thing not to be disputed about, of course." The remark resonates with quiet insistence: this work has an afterlife.

Whitman's body was displayed for public viewing on March 29, lying in an oak coffin installed in the rear parlor, surrounded by wreaths of flowers. Though certainly a stranger crowd than was customary for literary luminaries, thousands filed through Whitman's small house to pay homage to the departed poet.[133] Whitman would have been pleased with the diversity of the mourners, including working-class locals alongside the poet's intimates, disciples, and patrons. Traubel observed the "curious throng" with melancholy pleasure:

> The line grew longer and longer—it was silent, sympathetic, curious, expressive. It stretched out and up the street and then north through Fourth to the railroad—and it continued its reach and play for three hours till, at 1:50, we were compelled to stem and refuse it, in order to prepare for the cortege. Between twelve and one it took the simpler aspect of the laborers, off for their dinner hour. Letter carriers, policemen, railroadmen, ferrymen, school children, merchants—who was not included? I caught glimpses of tradesmen and familiar faces in all walks— men whom W. had known well and seen often and those to whom his kindness and gifts had added and stored precious affections.[134]

Onlookers lined Haddonfield Pike to observe the funeral procession; at Harleigh Cemetery, mourners covered the hillsides surrounding Whitman's tomb during the funeral service. The graveside scene was lively, the atmosphere reminiscent of a camp meeting as the crowd circulated on the hillside while bands played.[135] Refreshments were served as the ceremony was conducted on an elevated platform, from which Whitman's executors Bucke and Harned celebrated him as the poet of nature, humanity, and spirituality.[136] (Traubel elected not to speak, perhaps rendered mute by grief.) Harned delivered a eulogy on Whitman's literary immortality that was prophetically accurate: "that man is as indestructible as his Creator."[137] Finally enclosed in the crypt he had painstakingly designed, Whitman began the literary haunting that would dominate the American canon, having at last completed a memorial to contain the most obscure of his many specimens: that of the author.

NOTES

Abbreviations for Works of Walt Whitman

Corr.	*The Correspondence of Walt Whitman*, ed. Edwin Haviland Miller, 6 vols. (New York: New York University Press, 1961).
EPF	*The Collected Writings of Walt Whitman: The Early Poems and Fiction*, ed. Thomas L. Brasher (New York: New York University Press, 1963).
Journ.	*The Collected Writings of Walt Whitman: The Journalism 1834–1846*, edited by Herbert Bergmann, Douglas A. Noverr, and Edward J. Recchia (New York: Peter Lang, 1998).
LG 1855	*Leaves of Grass* (New York: 1855).
LG 1856	*Leaves of Grass* (Brooklyn, NY: Fowler and Wells, 1856).
LG 1860	*Leaves of Grass* (Boston: Thayer and Eldridge, 1860–61).
LG 1867	*Leaves of Grass* (New York: W. E. Chapin, 1867).
LG 1881	*Leaves of Grass* (Boston: James R. Osgood, 1881–82).
LG 1891	*Leaves of Grass* (Philadelphia: McKay, 1891–92).
MDW	*Memoranda during the War* (Camden, NY: Author's publication, 1875–76).
NUPM	*Notebooks and Unpublished Prose Manuscripts*, ed. Edward F. Grier (New York: New York University Press, 1984).
PP	*Walt Whitman: Poetry and Prose* (New York: Penguin Library of America, 1982).
PW	*Prose Works, 1892: Volume 2, Collected and Other Prose*, ed. Floyd Stovall (New York: New York University Press, 1964).
SD	*Specimen Days*, in *Complete Prose Works* (Philadelphia: David McKay, 1892).

Introduction

1. Walt Whitman to Louisa Van Velsor Whitman, September 18, 1863. Available on *The Walt Whitman Archive* (www.whitmanarchive.org).

2. *Oxford English Dictionary*.

3. Walt Whitman, *Memoranda during the War* (Camden, NY: Author's publication, 1875–76), 27. Hereafter *MDW*. Available on *The Walt Whitman Archive*.

4. As chapter 1 will show, most nineteenth-century Americans perceived dissection as a desecration of the corpse that had dire consequences for the soul in the afterlife,

as it prevented holy resurrection. Yet autopsies were usually performed not for experimental purposes, but to ascertain the cause of death for those deemed important enough to investigate. Thus, as Kenneth Nystrom argues, dissection was seen as "a form of structural violence," whereas autopsy was not necessarily stigmatized as such. However, as chapter 5 discusses in detail, Whitman's autopsy was performed for medically experimental purposes (to remove and study the poet's brain), and as an act of celebrity worship overseen by the poet's disciple Horace Traubel. Thus, I argue that Whitman's autopsy occupies a threshold moment in the intersection of mourning and medicine. See Kenneth C. Nystrom, "The Bioarchaeology of Structural Violence and Dissection in the 19th-Century United States," *American Anthropologist* 116, no. 4 (2014): 1–15.

5. Walt Whitman to Ralph Waldo Emerson, January 17, 1863. Available on *The Walt Whitman Archive.*

6. Whitman to Emerson, January 17, 1863.

7. Whitman to Emerson, January 17, 1863.

8. Walt Whitman, *Leaves of Grass* (Boston: Thayer and Eldridge, 1860–61), 373. Hereafter *LG 1860*. Available on *The Walt Whitman Archive.*

9. Whitman, *LG 1860*, 373–74.

10. Whitman, *MDW*, 28, 59.

11. Whitman to Emerson, January 17, 1863.

12. Whitman, *MDW,* 57; Walt Whitman, *Notebooks and Unpublished Prose Manuscripts*, ed. Edward F. Grier (New York: New York University Press, 1984), 2:504. Hereafter *NUPM*. My thinking on this point has evolved from Michel Foucault's theory of the medical gaze: "The clinical gaze is not that of an intellectual eye that is able to perceive the unalterable purity of essences beneath phenomena. It is a gaze of the concrete sensibility, a gaze that travels from body to body, and whose trajectory is situated in the space of sensible manifestation." Michel Foucault, *The Birth of the Clinic: An Archeology of Medical Perception* (New York: Routledge, 2003), 148.

13. Quoted by Ed Folsom, "Walt Whitman and the Civil War: Making Poetry Out of Pain, Grief, and Mass Death," *Abaton* 2 (Fall 2008): 13.

14. Whitman, *MDW*, 27.

15. Whitman, *MDW*, 48.

16. Walt Whitman to Louisa Van Velsor Whitman, September 18, 1863.

17. Whitman, *MDW*, 51.

18. See Drew Gilpin Faust, *This Republic of Suffering: Death and the American Civil War* (New York: Alfred A. Knopf, 2008), 6–10.

19. Faust, *Republic of Suffering*, 6–8; Mark S. Schantz, *Awaiting the Heavenly Country: The Civil War and America's Culture of Death* (Ithaca, NY: Cornell University Press, 2008), 89.

20. Ann Douglas, *The Feminization of American Culture* (New York: Alfred A. Knopf, 1977), 261.

21. Most recently, Adam C. Bradford has compellingly argued that Whitman's democratic ideal and spiritual vision depended on communal mourning rituals and the material proximity of the dead. Adam C. Bradford, *Communities of Death: Whitman, Poe, and the American Culture of Mourning* (Columbia: University of Missouri Press, 2014). Harold Aspiz's exhaustive study of death and the afterlife in Whit-

man's poetry has also been invaluable to this project. Harold Aspiz, *So Long! Walt Whitman's Poetry of Death* (Tuscaloosa: University of Alabama Press, 2004). See also William J. Scheick, "Death and the Afterlife," in *A Companion to Walt Whitman* (Malden, MA: Blackwell, 2006), 325–40. For an overview of sentimental mourning rituals, see Mary Louise Kete, *Sentimental Collaborations: Mourning and Middle-Class Identity in Nineteenth-Century America* (Durham, NC: Duke University Press, 2000); and Dana Luciano, *Arranging Grief: Sacred Time and the Body in Nineteenth-Century America* (New York: New York University Press, 2007). On the rise of spiritualism, see Molly McGarry, *Ghosts of Futures Past: Spiritualism and the Cultural Politics of Nineteenth-Century America* (Berkeley: University of California Press, 2008). On shifting attitudes toward the human cadaver, see Gary Laderman, *The Sacred Remains: American Attitudes toward Death, 1799–1883* (New Haven, CT: Yale University Press, 1996).

22. J. David Hacker, "A Census-Based Count of the Civil War Dead," *Civil War History* 57, no. 4 (2011): 310.

23. McGarry, *Ghosts of Futures Past*, 22–23.

24. Bradford, *Communities of Death*, 121.

25. Laderman, *Sacred Remains*, 27–28.

26. Laderman, *Sacred Remains*, 29.

27. Laderman, *Sacred Remains*, 35.

28. Laderman, *Sacred Remains*, 36.

29. Nicole Keller Day, "Death and Dying," in *Civil War America: A Social and Cultural History*, ed. Maggi M. Morehouse and Zoe Trodd (London: Routledge, 2013), 277. Embalming was, however, used by anatomists to preserve cadavers for dissection prior to the war, enhancing its negative stigma (Faust, *Republic of Suffering*, 92).

30. Walt Whitman, *Leaves of Grass* (Brooklyn: Fowler & Wells, 1856), 333. Hereafter *LG 1856*. See Aspiz, *So Long!* 78–88.

31. Whitman, *LG 1856*, 333.

32. Laderman, *Sacred Remains*, 76.

33. Walt Whitman, *Leaves of Grass* (New York, 1855), 37. Hereafter *LG 1855*. Available on *The Walt Whitman Archive*.

34. L. M. Sargent, *Dealings with the Dead, by a Sexton of the Old School*, vol. 1 (Boston: Dutton and Wentworth, 1856), 45–46; Laderman, *Sacred Remains*, 76.

35. Ralph Waldo Emerson, *Emerson: Essays and Lectures* (New York: Penguin Literary Classics, 1983), 194.

36. Ralph Waldo Emerson, *The Journals and Miscellaneous Notebooks of Ralph Waldo Emerson*, ed. Albert J. von Frank, vol. 4 (Columbia: University of Missouri Press, 1989), 7.

37. Laderman, *Sacred Remains*, 74.

38. Whitman, *MDW*, 48.

39. Whitman to Emerson, January 17, 1863.

40. Quoted in McGarry, *Ghosts of Futures Past*, 17.

41. Nicholas Abraham and Maria Torok, *The Shell and the Kernel: Renewals of Psychoanalysis*, ed. and trans. Nicholas T. Rand (Chicago: University of Chicago Press, 1994), 128; Whitman, *LG 1855*, 17.

42. Whitman, *MDW*, 57.

43. Walt Whitman, *Leaves of Grass* (Philadelphia: McKay, 1891–92), 452. Hereafter *LG 1891*. Available on *The Walt Whitman Archive*. Adam Bradford describes the first edition of *Leaves of Grass* as a "death-defying cryptext." His reading has been invaluable to mine: "Whitman's 1855 *Leaves of Grass* created what was, in essence, a death-defying cryptext. It sought to serve as a talisman, medium, and repository that could not only house a literary corpus, but also "enliven" that corpus in the reader's presence, ensuring a perpetual connection and communication between reader and Whitman" (Bradford, *Communities of Death*, 115).

44. Abraham and Torok, *Shell and the Kernel*, 140–41.

45. Sigmund Freud, "Mourning and Melancholia," in *On Murder, Mourning and Melancholia,* trans. Shaun Whiteside (New York: Penguin Modern Classics, 2005), 203–4. Hereafter "MM." See also Christopher Peterson, *Kindred Specters: Death, Mourning and American Affinity* (Minneapolis: University of Minnesota Press, 2007), 18. On Freud and Whitman, see Amy Parsons, "Desire, Forgetting and the Future: Walt Whitman's Civil War," *Arizona Quarterly* 71, no. 3 (2015): 101–2.

46. Freud, "MM," 204.

47. Sigmund Freud to Ludwig Binswanger, April 11, 1929, in *Letters of Sigmund Freud, 1873–1939*, ed. E. L. Freud, trans. Thomas Roberts (New York: Other, 2003), 386.

48. On this point I am indebted to Rose Lucas, "'Lines of Confusion, Stones of Emptiness': The Place of Mourning in Gwen Harwood's 'Herongate,'" *Southerly* 65, no. 1 (2005): 146–70.

49. Whitman to Emerson, January 17, 1863.

50. Whitman, *MDW*, 5; Jacques Derrida, *The Work of Mourning*, ed. Pascale-Anne Brault and Michael Naas (Chicago: University of Chicago Press, 2001), 159.

51. Derrida, *Work of Mourning*, 159; Jacques Derrida, *Memoires pour Paul de Man*, trans. Cecile Lindsay, Jonathan Culler, Eduardo Cadava, and Peggy Kamuf (New York: Columbia University Press, 1989), 144.

52. Derrida, *Work of Mourning*, 41–42.

53. Michael Nas, *Derrida from Now on* (New York: Fordham University Press, 2008), 170.

54. Silas Weir Mitchell, *Injuries of Nerves and Their Consequences* (Philadelphia: Lippincott, 1872), 348; Derrida, *Work of Mourning*, 41–42.

55. Whitman, *MDW*, 27.

56. Quoted in David S. Reynolds, *Walt Whitman's America: A Cultural Biography* (New York: Vintage, 1995), 411.

57. Whitman, *LG 1891*, 407.

58. Whitman, *LG 1860*, 342–43.

59. Whitman, *MDW*, 57.

60. See "This Compost" (Whitman, *LG 1856,* 205–6) and "A Twilight Song" (Whitman, *LG 1891*, 460–61).

61. Whitman, *LG 1891*, 347, 405.

62. Whitman, *MDW*, 5.

63. Whitman, *LG 1891*, 407.

64. Jacques Derrida, *On Touching—Jean-Luc Nancy,* trans. Christine Irizarry (Stanford, CA: Stanford University Press, 2005), 192; Jacques Derrida, *Aporias,* trans. Thomas Dutoit (Stanford, CA: Stanford University Press, 1993), 34.

65. Jacques Derrida, *Spectres of Marx: The State of Debt, The Work of Mourning, and The New International,* trans. Peggy Kamuf (New York: Routledge, 1994), 21.

66. J. Hillis Miller, "Absolute Mourning: It Is Jacques You Mourn for," in *Re-Reading Derrida: Perspectives on Mourning and Its Hospitalities*, ed. Tony Thwaites and Jude Seaboyer (New York: Lexington, 2013), 16.

67. Derrida, *Aporias*, 34.

68. Peter Coviello, *Intimacy in America: Dreams of Affiliation in Antebellum Literature* (Minneapolis: University of Minnesota Press, 2005), 127.

69. Whitman, *LG 1860*, *225*.

70. Whitman, *LG 1855*, 16; Whitman, *LG 1860*, 336; Whitman, *LG 1891*, 372.

71. Feldman reads Whitman's prose as "of a piece with the fragmenting and objectifying force of the museum's displays. The war along with Whitman's particular mode of representation turned the bodies of soldiers into objects. . . . Some of the order Whitman attains in his representation of the hospitals is produced by leaching the body of its individuality and condensing wounded soldiers into types, specimens, and cases." Mark Feldman, "Remembering a Convulsive War: Whitman's *Memoranda During the War* and the Therapeutics of Display," *Walt Whitman Quarterly Review* 23 (Summer 2005): 15–16. Conversely, Katherine Kinney articulates a fundamental distinction between Whitman's literary treatment of wounded soldiers and the museum aesthetic, which she links with the grotesque: "The withering flesh and the objective display of the museum mark precisely this loss of sensuality—a loss countered by the 'bodily excess' of Whitman's war prose in which injury and his response to it, including writing, remain sensuous human activity." Katherine Kinney, "Making Capital: War, Labor, and Whitman in Washington, D.C.," in *Breaking Bounds: Whitman and American Cultural Studies*, ed. Betsy Erkkila and Jay Grossman (New York: Oxford University Press, 1996), 174.

72. Adam C. Bradford, "Re-Collecting Soldiers: Walt Whitman and the Appreciation of Human Value," *Walt Whitman Quarterly Review* 27 (2010): 134.

73. Aspiz, *So Long!*

74. Robert Leigh Davis, *Whitman and the Romance of Medicine* (Berkeley: University of California Press, 1997), 6.

75. M. Wynne Thomas, *The Lunar Light of Whitman's Poetry* (Cambridge, MA: Harvard University Press, 1987), 278.

76. Martin T. Buinicki, *Walt Whitman's Reconstruction: Poetry and Publishing between Memory and History* (Iowa City: University of Iowa Press, 2011), 6.

77. Advertisement for "Manly Health" in *New York Atlas*, September 12, 1858. Quoted in Zachary Turpin, "Introduction to Walt Whitman's 'Manly Health and Training,'" *Walt Whitman Quarterly Review* 33 (2016): 155.

78. Turpin, "introduction," 148. The treatise invites comparison with Silas Weir Mitchell's *Wear and Tear, or Hints for the Overworked* (1871), touched on briefly in chapter 3, which prescribed gender-specific remedies, including a high-fat diet similar to Whitman's in "Manly Health," to combat the cerebral exhaustion brought on by modern urban life.

79. I am indebted to the following scholarship on Whitman and science: Harold Aspiz, "Science and Pseudoscience," in *A Companion to Walt Whitman*, ed. Donald D. Kummings (Oxford: Blackwell, 2009), 216–33; Joseph Beaver, *Walt Whitman: Poet of*

Science (New York: Octagon Books, 1974); Maria Farland, "Decomposing City: Walt Whitman's New York and the Science of Life and Death," *English Literary History* 74 (2007): 799–827; David S. Reynolds, "Earth, Body, Soul: Science and Religion," in *Walt Whitman's America: A Cultural Biography* (New York: Vintage, 1995), 235–79; Robert J. Scholnick, "'The Password Primeval': Whitman's Use of Science in 'Song of Myself,'" in *Studies in the American Renaissance 1986*, ed. Joel Myerson (Charlottesville: University of Virginia Press, 1886), 385–425; and Robert J. Scholnick, "Science," in *Walt Whitman: An Encyclopedia*, ed. J. R. LeMaster and Donald D. Kummings (New York: Garland, 1998), 616–19.

80. Walt Whitman, *Specimen Days*, in *Complete Prose Works* (Philadelphia: David McKay, 1892), 8. Hereafter *SD*. Available on *The Walt Whitman Archive.*

81. George Hutchinson and David Drews, "*Specimen Days,*" in *Walt Whitman: An Encyclopedia*, ed. J. R. LeMaster and Donald D. Kummings. Sourced from *The Walt Whitman Archive.*

82. Whitman, *SD,* 8.

83. This is not to suggest that *Specimen Days* is not chronological in organization (it is), but rather that Whitman's conception of "strange, unloosen'd, wondrous time" reflects nineteenth-century conceptions of fluid temporality in keeping with Dana Luciano's argument that "the altered flow of time experienced by mourners could be understood as a version of sacred time, the regenerative mode that transcended ordinary time in a ritual revisiting of origins" (Luciano, *Arranging Grief*, 7).

84. Walt Whitman, *Leaves of Grass* (New York: W. E. Chapin, 1867), 3b. Hereafter *LG 1867*.

85. Walter Benjamin, *The Arcades Project*, ed. Rolf Tiedemann, trans. Howard Eiland and Kevin McLaughlin (Cambridge, MA: Harvard University Press, 1999), 14.

86. Elizabeth Keeney, *The Botanizers: Amateur Scientists in Nineteenth-Century America* (Chapel Hill: University of North Carolina Press, 2011), 1.

87. C. Jeffrey, *An Introduction to Plant Taxonomy* (Cambridge: Cambridge University Press, 1982), 31.

88. As Betsy Erkkila observes, Whitman substantively altered the notebooks before publishing them as *Memoranda* and later *Specimen Days.* See Erkkila, *Whitman: The Political Poet* (Oxford: Oxford University Press, 1989), 207.

89. Whitman's specimens are described as "odd," "unworld[ly]," "disinterested," "veiled and abstracted" (Whitman, *MDW*, 27).

90. Thomas Hallock, "Male Pleasure and the Genders of Eighteenth-Century Botanic Exchange: A Garden Tour," *William and Mary Quarterly* 62, no. 4 (2005): 697–718.

91. Elizabeth A. Petrino, "'Silent Eloquence': The Social Codification of Floral Metaphors in the Poems of Frances Sargent Osgood and Emily Dickinson," *Legacy* 15, no. 2 (1998): 139.

92. Judith Farr with Louise Carter, *The Gardens of Emily Dickinson* (Cambridge, MA: Harvard University Press, 2004), 84. See also Christine Gerhardt, *A Place for Humility: Whitman, Dickinson and the Natural World* (Iowa City: University of Iowa Press, 2014), 25.

93. Petrino, "Silent Eloquence," 140.

94. Emily Dickinson, *The Collected Poems,* ed. Thomas H. Johnson (London: Faber and Faber, 1975), 49.

95. Farr and Carter, *Gardens of Emily Dickinson*, 84.

96. Ruskin quoted in Farr and Carter, *Gardens of Emily Dickinson*, 85.

97. This was the third volume of *The Narrative of the Voyages of H. M. Ships Adventure and Beagle,* edited by Captain Robert Fitzroy and published in three volumes in 1839. In this form, Darwin's text holds the subsidiary title *Journal and Remarks.* The title subsequently changed four times; today it is referred to as *The Voyage of the Beagle.* See R. B. Freeman, *The Works of Charles Darwin: An Annotated Bibliographical Handlist* (Dawson, UK: Folkstone, 1977). Sourced from John van Wyhe, ed., 2002–, *The Complete Work of Charles Darwin Online* (http://darwin-online.org.uk/).

98. Whitman, *MDW,* 4; Richard Keynes, ed., *Charles Darwin's Zoology Notes & Specimen Lists from H.M.S. Beagle* (Cambridge: Cambridge University Press, 2000), 317.

99. Whitman, *MDW,* 3.

100. Whitman, *MDW,* 29–30.

101. Whitman, *MDW,* 17.

102. Peter Coviello, ed., "Preface: Whitman at War," in *Memoranda during the War,* by Walt Whitman (Oxford: Oxford University Press, 2004), xxix.

103. As chapter 3 details, my argument is influenced by Foucault's thesis that during the later nineteenth century, the "homosexual became a species." Michel Foucault, *The History of Sexuality,* vol. 1 (New York: Vintage, 1990), 43.

104. For further analysis of Whitman's reading of Darwin, see Aspiz, "Science and Pseudoscience," 216–33; Beaver, *Walt Whitman: Poet of Science,* 105–30; Reynolds, *Walt Whitman's America,* 246, 481–82.

105. The first inclusion of the term "specimen" in *Leaves of Grass* occurred in an anonymous review of the 1855 edition that was reprinted in the 1856 edition: "We have before us one of the most extraordinary specimens of Yankee intelligence and American eccentricity in authorship, it is possible to conceive" (Whitman, *LG 1856,* 359; originally printed in *The London Weekly Dispatch*, March 9, 1856). It is not difficult to see why this review was embraced by the self-promoting author. The word "specimen" appears three times in the 1860 edition, twice in 1867 and 1881, and three times in 1892 (once in "A Backward Glance o'er Travel'd Roads"). It is used numerous times in *Memoranda during the War* and *Specimen Days*.

106. Whitman, *SD,* 177.

107. Horace Traubel, *With Walt Whitman in Camden,* vol. 8, ed. Jeanne Chapman and Robert MacIsaac (Oregon House, CA: W.L. Bentley, 1996), 454. Available on *The Walt Whitman Archive.*

108. Whitman, *SD,* 197–98.

109. Whitman, *LG 1860*, 225.

110. Horace Traubel, *With Walt Whitman in Camden,* vol. 6 (Carbondale: Southern Illinois University Press, 1982), 129. Available on *The Walt Whitman Archive.*

111. Charles Darwin, *The Origin of Species*, ed. Gillian Beer (Oxford: Oxford University Press, 1996), 252.

112. Whitman, *MDW,* 4.

113. Walt Whitman, *The Correspondence of Walt Whitman*, ed. Edwin Haviland Miller, 6 vols. (New York: New York University Press, 1961), 1:81. Hereafter *Corr.*

114. *Oxford English Dictionary.*

115. Whitman, *Corr.*, 1:81.

116. Whitman, *Corr.*, 1:81.

117. Whitman, *Corr.*, 1:81.

118. Whitman, *MDW*, 69.

119. Whitman, *MDW*, 57.

120. Darwin, *Origin of Species*, 257.

121. Whitman, *MDW*, 27.

122. Whitman, *MDW*, 27.

123. Darwin, *Origin of Species*, 252, 254.

124. Whitman, *Corr.*, 1:81.

125. Whitman, *MDW*, 57.

126. Whitman, *LG 1855*, 710.

127. Whitman, *MDW*, 15.

128. Whitman, *NUPM*, 2:504.

129. Whitman, *LG 1891*, 372.

130. Whitman, *NUPM*, 2:504; Mitchell, *Injuries of Nerves*, 348.

131. *LG 1867*, 11b.

132. Whitman, *LG 1891*, 247.

133. Technically, the "deathbed edition" is not actually a distinct edition but is a reprinting of the 1881–82 edition with "annexes." See *The Walt Whitman Archive* for publication history.

134. Horace Traubel, *With Walt Whitman in Camden*, vol. 9, ed. Jeanne Chapman and Robert MacIsaac (Oregon House, CA: W.L. Bentley, 1996), 605.

Chapter One

1. Whitman, *LG 1855*, 14.

2. Whitman, *LG 1855*, 16.

3. Walt Whitman, *The Collected Writings of Walt Whitman: The Early Poems and Fiction*, ed. Thomas L. Brasher (New York: New York University Press, 1963), 88. Hereafter *EPF*. As Ed Folsom and Kenneth M. Price observe, "The Tomb Blossoms" also marked Whitman's "first appearance" in a published book, *Voices from the Press* (1850). Folsom and Price, *Re-Scripting Walt Whitman: An Introduction to His Life and Work* (Malden, MA: Blackwell, 2005), 16.

4. Whitman, *EPF*, 88–89.

5. Whitman, *EPF*, 89–90.

6. Whitman, *EPF*, 91.

7. Whitman, *EPF*, 92.

8. Whitman, *EPF*, 92.

9. Whitman, *EPF*, 93.

10. Whitman, *EPF*, 93.

11. Whitman, *EPF*, 93. Harold Aspiz links this passage with the established traditions of consolation literature (*So Long!* 15).

12. Whitman, *LG 1855*, 54.

13. Whitman, *LG 1855, 16.*

14. As Kenneth M. Price writes, "In the poem ultimately titled 'Song of Myself,' . . . and in many other passages, Whitman explores the idea of death or destruction as both ruination and renewal, a process in which things are broken down to an elemental condition for whatever comes next." Kenneth M. Price, "Debris, Creative Scatter, and the Challenges of Editing Whitman," in *Walt Whitman: Where the Future Becomes Present,* ed. David Haven Blake and Michael Robertson (Iowa City: University of Iowa Press, 2008).

15. Whitman, *LG 1860,* 342–43.

16. All poems were originally untitled in the 1855 edition. To avoid confusion, I refer to this poem by its final title, "Song of Myself."

17. Whitman, *MDW,* 5.

18. Walt Whitman, *The Collected Writings of Walt Whitman: The Journalism 1834–1846,* ed. Herbert Bergmann, Douglas A. Noverr, and Edward J. Recchia (New York: Peter Lang, 1998), 1: 92. Hereafter *Journ.*

19. Whitman, *Journ.,* 1: 92.

20. Whitman, *Journ.,* 1: 92.

21. Peter N. Stearns, "Anatomy and Punishment in Late Eighteenth-Century New York," *Journal of Social History* 22, no. 3 (1989): 510.

22. As Michael Sappol has established, "medical education in America expanded rapidly in the nineteenth-century, from four schools in 1800 to more than 160 in 1900." Michael Sappol, *A Traffic of Dead Bodies: Anatomy and Embodied Social Identity in Nineteenth-Century America* (Princeton, NJ: Princeton University Press, 2002), 2.

23. Whitman, *NUPM,* 6:2122–23.

24. Sappol, *Traffic of Dead Bodies,* 4.

25. Whitman, *LG 1855,* 78.

26. Sappol, *Traffic of Dead Bodies,* 4–5; Aaron D. Tward and Hugh A. Patterson, "From Grave Robbing to Gifting: Cadaver Supply in the United States," *Journal of the American Medical Association* 287, no. 9 (2002): 1183.

27. Walt Whitman, "Broadway Hospital," *New York Leader,* March 22, 1862. Available on *The Walt Whitman Archive.* Broadway Hospital was also known as New York Hospital.

28. Sappol, *Traffic of Dead Bodies,* 108–9.

29. Bess Lovejoy, "The Gory New York City Riot That Shaped American Medicine," *The Smithsonian,* June 17, 2014, http://www.smithsonianmag.com/history/gory-new-york-city-riot-shaped-american-medicine-180951766/.

30. See Suzanne M. Shultz, *Body Snatching: The Robbing of Graves for the Education of Physicians in Early Nineteenth-Century America* (Jefferson, NC: McFarland, 1992), 79–80.

31. See Edward Robb Ellis, *The Epic of New York City: A Narrative History* (New York: Carrol & Graf, 2004), 181–83.

32. Tward and Patterson, "From Grave Robbing to Gifting," 1183.

33. Stearns, "Anatomy and Punishment," 512.

34. Sappol, *Traffic of Dead Bodies,* 10–11.

35. Shultz, *Body Snatching,* ix.

36. Frederick C. Waite, "Grave Robbing in New England," *Bulletin of the Medical Library Association* 33, no. 3 (1945): 281.

37. Carolyn Marvin, "The Body of the Text: Literacy's Corporeal Constant," *The Quarterly Journal of Speech* 80, no. 2 (1994): 138. For example, as Hubert Cole writes, "A young doctor, John Collins Warren, studying at the United Hospitals in the Borough, wrote home to Boston in December, 1799: 'Dissection is carried on in style: twelve or fifteen bodies in a room; the young men work on them in different ways. The people called resurrection men supply us abundantly.'" Hubert Cole, *Things for the Surgeon: A History of the Body-Snatchers* (London: Heineman, 1964), 11.

38. As Marvin explains, "The Christian doctrine of the physical resurrection of the corpse was one source of tension between knowledge derived from analyzing texts, a kind of dissection, and from dissecting bodies, a kind of analysis. Medical practice of all kinds was suspect among the highest ranks of the clergy and accompanied by specific doctrinal disapproval of anatomical dissection. Traditional beliefs about the essential connection between the body and the soul for a specified period following death, along with traditional customs that placed the corpse at the center of funerary rites, stoked popular fears of medical grave-robbing, and nourished popular resistance to dissection well into the nineteenth century" (Marvin, "Body of the Text," 138).

39. Justin Kaplan, *Walt Whitman: A Life* (New York: Simon & Schuster, 1980), 75–76.

40. Sappol, *Traffic of Dead Bodies*, 14.

41. Shultz, *Body Snatching*, 44.

42. Waite, "Grave Robbing in New England," 279–80.

43. Waite, "Grave Robbing in New England," 277–80.

44. Whitman, *EPF*, 92.

45. A disproportionate number of anatomical subjects were African American, Indian, or Irish. Female prostitutes were also a primary source of anatomical material (Sappol, *Traffic of Dead Bodies*, 2–5).

46. David C. Humphrey, "Dissection and Discrimination: The Social Origins of Cadavers in America, 1760–1915," *Bulletin of the New York Academy of Medicine* 49, no. 9 (1973): 822.

47. The legend of the name dates back to Judas Iscariot, who as an act of repentance returned to the temple the thirty pieces of silver he was paid for the betrayal of Christ. The priests decreed that the blood money could not enter the treasury, and instead used it to purchase a plot for the burial of strangers who died in the city of Jerusalem. They acquired an abandoned field, formerly the site of a pottery, which was known locally as potter's field (Waite, "Grave Robbing in New England," 279).

48. Thomas S. Sozinsky, "Grave Robbing and Dissection," *Penn Monthly* 10 (1879): 217.

49. Humphrey, "Dissection and Discrimination," 822.

50. Edward C. Halperin, "A Glimpse of Our Past: The Poor, the Black and the Marginalized as the Source of Cadavers in United States Anatomical Education," *Clinical Anatomy* 20 (2007): 493.

51. Halperin, "Glimpse of Our Past," 493.

52. Humphrey, "Dissection and Discrimination," 820.

53. Shultz, *Body Snatching*, 124.

54. Humphrey, "Dissection and Discrimination," 820.

55. Humphrey, "Dissection and Discrimination," 820.

56. Humphrey, "Dissection and Discrimination," 819.

57. Humphrey, "Dissection and Discrimination," 819.
58. Shultz, *Body Snatching*, 27–28.
59. Sappol, *Traffic of Dead Bodies*, 128.
60. Sappol, *Traffic of Dead Bodies*, 132.
61. Humphrey, "Dissection and Discrimination," 824.
62. Sappol, *Traffic of Dead Bodies*, 4.
63. See Faust, *Republic of Suffering*, 6–8; Schantz, *Awaiting the Heavenly Country*, 89.
64. Sappol, *Traffic of Dead Bodies*, 2.
65. Whitman, *LG 1855*, 17.
66. Whitman, *EPF*, 92.
67. Whitman, *LG 1855*, 14.
68. See Aspiz, "Science and Pseudoscience," 216–33.
69. Whitman, *LG 1855*, vii.
70. On "skin trades," see Sappol, *Traffic of Dead Bodies*, 4; Whitman, *LG 1855*, 80.
71. Whitman, *LG 1855*, 80–81.
72. Whitman, *LG 1855*, 82.
73. Whitman, *LG 1855*, 82.
74. For further analysis of Whitman's ambivalent treatment of race, see Ed Folsom, "Lucifer and Ethiopia: Whitman, Race, and Poetics Before the Civil War and After," in *A Historical Guide to Walt Whitman*, ed. David S. Reynolds (Oxford: Oxford University Press, 2000), 45–96; Isaac Gewirtz "I Am with [Some of] You," in *I Am with You: Walt Whitman's "Leaves of Grass" 1855–2005* (New York: New York Public Library, 2005), 10–46; Reynolds, *Walt Whitman's America*, 471–72; and Ivy G. Wilson, ed., *Whitman Noir: Black America and the Good Grey Poet* (Iowa City: University of Iowa Press, 2014).
75. Quoted in Gewirtz, "I Am with [Some of] You," 36.
76. Whitman, *LG 1855*, 26.
77. Whitman, *LG 1855*, 24.
78. Whitman, *LG 1855*, 16.
79. Laderman, *Sacred Remains*, 42.
80. Maria Farland, "Decomposing City: Walt Whitman's New York and the Science of Life and Death," *English Literary History* 74 (2007): 807.
81. John H. Griscom, *The Sanitary Condition of the Laboring Population of New York with Suggestions for Improvement* (New York: Harper, 1845), 53.
82. Wilson Flagg, *Mount Auburn: Its Scenes, Its Beauties, and Its Lessons* (Boston: James Munroe, 1861), 321.
83. Whitman, *Journ.*, 1:309.
84. Farland, "Decomposing City," 801.
85. Whitman, *Journ.*, 1:309.
86. Whitman, *LG 1855*, 55.
87. Whitman, *LG 1855*, 55, 27.
88. Whitman, *LG 1867*, 306.
89. Whitman, *LG 1855*, 16.
90. Judith Butler, *Antigone's Claim: Kinship between Life and Death* (New York: Columbia University Press, 2000), 47.
91. Butler, *Antigone's Claim*, 13–14.
92. Whitman, *EPF*, 92.

93. Whitman, *LG 1855,* 17.

94. Whitman, *LG 1855,* 54.

95. Whitman, *LG 1855,* 16.

96. Whitman, *LG 1855,* 16.

97. Freud, "MM," 204.

98. Abraham and Torok, *Shell and the Kernel,* 118, 120–23. The term "exquisite corpse" was appropriated by Torok from Andre Breton's surrealist word game. On their work with Holocaust survivors, see Lawrence Johnson, "Cryptonymic Secretions: On the Kindness of Strangers," in *Re-Reading Derrida: Perspectives on Mourning and Its Hospitalities,* ed. Judith Seaboyer and Tony Thwaites (New York: Lexington Books, 2013), 126.

99. Abraham and Torok, *Shell and the Kernel,* 125–26.

100. Abraham and Torok, *Shell and the Kernel,* 128.

101. Abraham and Torok, *Shell and the Kernel,* 126.

102. *Oxford English Dictionary.*

103. Whitman, *Journ.,* 1:238.

104. Francis D. Allen, *Documents and Facts, Showing the Fatal Effects of Interments in Populous Cities* (New York: New York Common Council, 1822), iii.

105. Laderman, *Sacred Remains,* 70.

106. Farland, "Decomposing City, 825.

107. Farland, "Decomposing City," 808.

108. Desirée Henderson, "'What Is the Grass?': The Roots of Walt Whitman's Cemetery Meditation," *Walt Whitman Quarterly Review* 25, no. 3 (2008): 95.

109. Quoted in Reynolds, *Walt Whitman's America,* 240.

110. Justus Liebig, *Chemistry in Its Application to Agriculture and Physiology* (Cambridge, MA: John Owen, 1842), 297.

111. Liebig, *Chemistry,* 400.

112. Whitman, *LG 1855,* 13; Reynolds, *Walt Whitman's America,* 240.

113. Whitman, *LG 1855,* 17.

114. Whitman, *LG 1855,* 51.

115. Ralph Waldo Emerson, *Emerson: Essays and Lectures* (New York: Penguin Literary Classics, 1983), 30.

116. Walt Whitman, *An American Primer,* ed. Horace Traubel (Boston: Small, Maynard, 1904), 21.

117. Whitman, *LG 1855,* 55.

118. Whitman, *LG 1855,* 26.

119. Walt Whitman, *Uncollected Poetry and Prose,* ed. Emory Holloway (Garden City, NY: Doubleday, 1921), 1:254.

120. James F. Johnston, *The Chemistry of Common Life* (New York: D. Appleton, 1855), 13. On this point I am again indebted to Farland's analysis of correlations between Johnston and Whitman.

121. Whitman, *LG 1855,* 16.

122. Whitman, *LG 1855,* 16. Farland suggests that antebellum theories of biochemistry offer insights into the "bizarre formulations" in *Leaves of Grass* "concerning the poetic voice and the tongue" (Farland, "Decomposing City," 819).

123. Whitman, *LG 1855,* 49.

124. Whitman, *MDW*, 56–57.
125. Whitman, *LG 1855*, 38.
126. Whitman, *LG 1855*, 39.
127. Whitman, *LG 1855*, 39.
128. McGarry, *Ghosts of Futures Past*, 60.
129. McGarry, *Ghosts of Futures Past*, 60.
130. McGarry, *Ghosts of Futures Past*, 60.
131. Whitman, *LG 1855*, vii.
132. Reynolds, *Walt Whitman's America*, 263.
133. Between the publication of the first and second editions, Whitman became fascinated with the phenomenon of spiritualism, and attempted for an entire year to train himself, modelling his practice after the famous medium Cora Hatch (McGarry, *Ghosts of Futures Past*, 170). Whitman eventually became disenchanted with spiritualism, or at least with some of its proponents. On December 16, 1874, he responded to the request of David Goodman Croly, the editor of the *New York Daily Graphic*, that he write a piece on spiritualism: "I am neither disposed nor able to write anything about this so-called Spiritualism. . . . It seems to me nearly altogether a poor, cheap, crude humbug." Jerome Loving, *Walt Whitman: The Song of Himself* (Berkeley: University of California Press, 1999), 353.
134. Jacques Derrida, quoted by Anne Dufourmantelle, from an unpublished seminar conducted in Paris, January 1996. "Invitation," in *Of Hospitality*, trans. Rachel Bowlby (Stanford, CA: Stanford University Press, 2000), 144.
135. Dufourmantelle, "Invitation," 4.
136. Abraham and Torok, *Shell and the Kernel*, 126.
137. Butler, *Antigone's Claim*, 78.
138. Michael Moon, "Memorial Rags," in *Professions of Desire: Lesbian and Gay Studies in Literature*, ed. George Haggerty and Bonnie Zimmerman (New York: MLA, 1995), 451.
139. Abraham and Torok, *Shell and the Kernel*, 110.
140. Whitman, *LG 1855*, 16.
141. Whitman, *LG 1855*, 15.
142. Whitman, *LG 1855*, 15.
143. Abraham, and Torok, *Shell and the Kernel*, 116.
144. Whitman, *LG 1855*, 27.
145. My reading of Whitman's ecoeroticism is indebted to M. Jimmie Killingsworth's ecocritical analysis, *Walt Whitman and the Earth: A Study in Ecopoetics* (Iowa City: University of Iowa Press, 2004); and to Byrne S. Fone, *Masculine Landscapes: Walt Whitman and the Homoerotic Text* (Carbondale: Southern Illinois University Press, 1992).
146. Whitman, *LG 1855*, 28.
147. Whitman, *LG 1855*, 56.
148. Abraham and Torok, *Shell and the Kernel*, 137.
149. *Oxford English Dictionary.*
150. Whitman, *LG 1855*, 56.
151. Whitman, *LG 1855*, 56.
152. Whitman, *EPF*, 92.

153. James D. McCabe, Jr., *Lights and Shadows of New York Life; or, The Sights and Sensations of the Great City* (New York: National Publishing Company, 1872), 841.

154. Joann P Krieg, "Walt Whitman and the Prostitutes," *Literature and Medicine* 14, no. 1 (1995): 36–52.

155. Howard Markel, *An Anatomy of Addiction: Sigmund Freud, William Halstead and the Miracle Drug Cocaine* (New York: Pantheon Books, 2011), 39–41.

156. Folsom and Price, *Re-Scripting Walt Whitman,* 90.

157. Markel, *Anatomy of Addiction*, 41–42.

158. McCabe, *Lights and Shadows*, 839.

159. Walt Whitman, "New York Dissected," *Life Illustrated,* August 16, 1856, 125. Available on *The Walt Whitman Archive.*

160. As Christopher Freeburg writes, "Whitman was a racist, and he did subscribe to white supremacist ideas and attitudes. The most interesting of his remarks about blacks is contained in the following realization: 'Blacks can never be to me what whites are. Below all political relations, even the deepest, are still deeper, personal and emotional ones, the whites are my brothers & I love them.'" Christopher Freeburg, "Walt Whitman, James Weldon Johnson and the Violent Paradox of U.S. Progress," in *Whitman Noir,* ed. Wilson, 90.

161. Erkkila, *Whitman: The Political Poet*, 102.

162. Ivy G. Wilson, "Looking with a Queer Smile: Walt Whitman's Gaze and Black America," in *Whitman Noir,* ed. Wilson, ix.

163. Paradoxically, in contrast to his writings on racial inequality, in "Thoughts" (1867) Whitman writes, "Of Equality—As if it harm'd me, giving others the same chances and rights as myself—As if it were not indispensible to my own rights that others possess the same" (Whitman, *LG 1867,* 261).

164. Jonathan Strauss, *Human Remains: Medicine, Death, and Desire in Nineteenth-Century Paris* (New York: Fordham University Press, 2012), 165.

165. Hippolyte Mireur, *La Syphilis et la prostitution dans leurs rapports avec l'hygiene, la morale et la loi* (Paris: Masson, 1888); Strauss, *Human Remains*, 164.

166. Luke Mancuso, "*Leaves of Grass*, 1867 Edition," in *Walt Whitman: An Encyclopedia*, ed. J. R. LeMaster and Donald D. Kummings (New York: Garland, 1998). Available on *The Walt Whitman Archive.*

167. A draft of "The City Dead-House" appears in an 1862–63 notebook alongside notes from the field hospital at Falmouth, Virginia. Thomas Biggs Harned Collection of the Papers of Walt Whitman, Library of Congress. Hereafter, Harned Whitman Collection.

168. Folsom and Price, *Re-Scripting Walt Whitman,* 79.

169. Robert Roper, *Now the Drum of War: Walt Whitman and His Brothers in the Civil War* (New York: Walker, 2008), 118.

170. Folsom and Price, *Re-Scripting Walt Whitman,* 79; Roper, *Now the Drum of War*, 119.

171. In a letter to his mother dated December 29, 1862, Whitman states that he arrived in Falmouth on December 19 and returned to Washington on December 28. Whitman, *Corr.,* 1:59.

172. Whitman, 1862–63 notebook, Harned Whitman Collection.

173. Whitman, 1862–63 notebook, Harned Whitman Collection.

174. Whitman, 1862–63 notebook, Harned Whitman Collection.

175. David J. Eicher, *The Longest Night: A Military History of the Civil War* (New York: Simon & Schuster, 2001), 405.

176. Roper, *Now the Drum of War*, 145.

177. Whitman, *EPF*, 92.

178. Whitman, "City Photographs," *New York Leader*, March 22, 1862: [1]. Sourced from *The Walt Whitman Archive.*

179. Whitman, "City Photographs."

180. Roy Morris, Jr., *The Better Angel: Walt Whitman in the Civil War* (Oxford: Oxford University Press, 2000), 33.

181. Whitman, *LG 1855*, 21.

182. Whitman, *LG 1855*, 16.

183. Kenneth M. Price, *To Walt Whitman, America* (Chapel Hill: University of North Carolina Press, 2004), 5.

184. Roper, *Now the Drum of War*, 115.

185. Whitman, "City Photographs."

186. Whitman, *NUPM*, 2:504.

Chapter Two

1. Lacy House still stands, as do Whitman's catalpa trees. The property is now known as Chatham Manor. See Robert Schultz, "Comment: Letter from Virginia," *The Hudson Review* 68, no. 1 (2015): 12–15.

2. Whitman, *MDW*, 15, 8–9, 102.

3. Whitman, *Corr.*, 1:59.

4. Brinton had arrived five days earlier on December 14, 1862. John H. Brinton, *The Personal Memoirs of John H. Brinton, Major and Surgeon, 1861–65* (New York: Neale, 1914), 214. Hereafter *PM*.

5. Brinton, *PM*, 187.

6. Brinton, *PM*, 190.

7. See Schantz, *Awaiting the Heavenly Country*, 184. See also Timothy Sweet, *Traces of War: Poetry, Photography and the Crisis of Union* (Baltimore: Johns Hopkins University Press, 1990), 107–37; and Max Cavitch, *American Elegy: The Poetry of Mourning from the Puritans to Whitman* (Minneapolis: University of Minnesota Press, 2007), 239–40.

8. Whitman, *NUPM*, 2:504.

9. Caleb E. Finch, "Aging, Inflammation and the Body Electric," *Daedalus* 135 (2006): 68. See also Loving, *Walt Whitman: The Song of Himself*, 479.

10. Whitman, *MDW*, 27.

11. Oliver Wendell Holmes, "The Human Wheel, Its Spokes and Felloes," *Atlantic Monthly*, May 1863, 567.

12. Amputation was the Civil War's most common surgical procedure. More than 60,000 limbs were severed, more than during any other war in U.S. history. See Laurann Figg and Jane Farrell-Beck, "Amputation in the Civil War: Physical and Social Dimensions," *Journal of the History of Medicine and Allied Sciences* 48, no. 4 (1993): 454.

13. Guy R. Hasegawa, *Mending Broken Soldiers: The Union and Confederate Programs to Supply Artificial Limbs* (Carbondale: Southern Illinois University Press, 2012), 2.

14. Whitman, *MDW,* 69.

15. Whitman, *MDW,* 27; Walt Whitman, *Prose Works, 1892: Volume 2, Collect and Other Prose*, ed. Floyd Stovall (New York: New York University Press, 1964), 504. Hereafter *PW.*

16. *Oxford English Dictionary.*

17. Freud, "MM," 262.

18. Walt Whitman, *November Boughs,* in *Walt Whitman: Poetry and Prose* (New York: Penguin Library of America, 1982), 1214, emphasis mine. Hereafter *PP.*

19. Sigmund Freud, *Studies on Hysteria*, trans. and ed. James Strachey (New York: Basic Books, 1957), 221.

20. Brinton, *PM*, 180.

21. Brinton, *PM,* 186; Mark Seltzer, "Wound Culture: Trauma in the Pathological Public Sphere," *October* 80 (Spring 1997): 3.

22. Seltzer, "Wound Culture," 3; Nancy Cervetti, *S. Weir Mitchell: Philadelphia's Literary Physician, 1829–1914* (University Park: Pennsylvania State University Press, 2012), 81.

23. Brinton, *PM,* 189.

24. Seltzer, "Wound Culture," 3–4.

25. Whitman, *MDW,* 4–5.

26. Whitman, *MDW,* 4.

27. Whitman, *MDW,* 16.

28. Whitman, *MDW,* 3–4.

29. Sweet has discussed Whitman's use of synecdoche in *Memoranda* and *Drum-Taps*, "in which a single event or experience replaces the whole of the war" (Sweet, *Traces of War*, 48). Betsy Erkkila has identified Whitman's inclusivity with America's unofficial motto: *e pluribus unum* ("out of many one"). See Erkkila, *Whitman: The Political Poet*, 81–91.

30. Brinton, *PM,* 186. Kinney has discussed the influence of museum aesthetics on Whitman's war prose: "The crude example of the Army Medical Museum proved itself to be only a curiosity. It may perhaps have been an intermediate gesture, a necessary demonstration of the government's ability to control, order, and enumerate the wounded and the dead in the immediate aftermath of the war. But it offers at least a suggestive explanation for why Whitman would claim again and again that his war prose was without literary design. These memoranda are not formal systems, he insists, but haphazard, random jottings, without any mediating gesture of poetry or narrative" (Kinney, "Making Capital," 185). I agree that Whitman's specimens resist categorization (they are elusive, "otherworldly" beings, for all the poet's attempts to narrate their commonalities). I also read this as a response to the medical classification represented in the Army Medical Museum. However, Kinney's dismissal of the museum as a "crude" and fleeting "curiosity" is problematic, if for no other reason than the fact that the museum remains open to this day, and currently displays hundreds of Civil War specimens, including the remains of soldiers nursed by Whitman.

31. Whitman, *PW,* 504; Whitman, *MDW,* 28.

32. Whitman, *LG 1891,* 452.

33. Cathy Caruth, *Unclaimed Experience: Trauma, Narrative and History* (Baltimore: Johns Hopkins University Press, 1996), 4.

34. Sigmund Freud, *Beyond the Pleasure Principle,* in *On Metapsychology: The Theory of Psychoanalysis* (New York: Penguin, 1991), 289. Hereafter *BPP.*

35. Buinicki, *Walt Whitman's Reconstruction*, 52–53.

36. Whitman, *PP,* 1210. See Buinicki's analysis of this passage in relation to the problems of accurately representing traumatic memory (*Walt Whitman's Reconstruction,* 52–54).

37. Whitman, *MDW,* 4.

38. As Betsy Erkkila observes, Whitman substantively altered the notebooks before publishing them as *Memoranda* and later *Specimen Days* (*Whitman: The Political Poet,* 207).

39. Whitman, *MDW,* 4.

40. Whitman, *MDW,* 4.

41. See Robert Leigh Davis on the hospitals' liminality (*Whitman and the Romance of Medicine*, 6).

42. Whitman to Louisa Van Velsor Whitman, September 8, 1863 (*Corr.,* 1:145–46).

43. Whitman, *MDW,* 4.

44. Whitman, *MDW,* 4.

45. Sweet links the traumatic resonance of the notebooks "not so much with their words, but because they bear traces of the violated human body which cannot be represented and thus must fail to become part of the public record of war" (*Traces of War,* 48).

46. Walt Whitman, *Leaves of Grass* (Boston: James R. Osgood, 1881–82), 241. Hereafter *LG 1881.* Available on *The Walt Whitman Archive.*

47. Whitman, *MDW,* 4.

48. Feldman has also observed the "link between writing and wounding" in this passage ("Remembering a Convulsive War," 8).

49. Brinton, *PM,* 169.

50. Brinton, *PM,* 180.

51. Brinton, *PM,* 8.

52. Brinton, *PM,* 180–81.

53. Faust, *Republic of Suffering,* 95.

54. The move commenced in 1866, and the renovated museum opened to the public in 1867. It became an increasingly popular Washington landmark throughout the later nineteenth century. Michael Rhode, "An Army Museum or a National Collection? Shifting Interests and Fortunes at the National Museum of Health and Medicine," in *Medical Museums: Past, Present, Future,* ed. Samuel J. M. M. Alberti and Elizabeth Hallam (London: Royal College of Surgeons of England, 2013), 190.

55. Brinton, *PM,* 27.

56. Brinton, *PM,* 15.

57. The phrase "rejected members" is Whitman's: "One man is shot by a shell, both in the arm and leg—both are amputated—there lie the rejected members" (*MDW,* 15). The phrase "mutilated limbs" is Brinton's: "It was hard enough to be worked day and night in those great surgical emergencies, accompanying fierce and protracted battles, and it really seemed unjust to expect the rough preparation, necessary to preserve for the Museum, the mutilated limbs" (*PM,* 187).

58. Whitman, *MDW,* 6.

59. Brinton, *PM,* 91.

60. A total of 5,532 contract surgeons served in the Union Army. See the entry "Acting Assistant (Contract) Surgeons" in Glenna R. Schroeder-Lein's *Encyclopedia of Civil War Medicine* (New York: M.E. Sharp, 2008). "Early in the war, contract surgeons were heavily criticized. After the Battle of Antietam, newspapers complained that many doctors who had come to the battlefield were opportunists who performed unnecessary amputations because they wanted surgical experience, rather than to help the patients." However, Schroeder-Lein goes on to emphasize that, while criticism was justified in certain instances, "most short-term medical workers were competent and did good work" (295).

61. Brinton, *PM,* 66.

62. Julian John Chisholm, *A Manual of Military Surgery for the Use of the Surgeons in the Confederate Army*, 3rd ed. (Columbia, SC: Evans and Cogswell, 1864), 409.

63. Whitman, *MDW,* 9–10.

64. As Feldman has observed, in Whitman's context the word "convulsion" "contains contradictory and divisive ideas. The prefix *con-* means together, while the Latin verb *vellare* means to pull or tear" (Feldman, "Remembering a Convulsive War," 1).

65. Freud, "MM," 262.

66. Freud, "MM," 253.

67. Traubel, *With Walt Whitman in Camden*, vol. 3 (New York: Mitchell Kinnerley, 1914), 581–82.

68. On this point, I am indebted to Moon's essay "Memorial Rags" (451).

69. Whitman, *MDW*, 10.

70. This event occurred around December 15, as the wounded were evacuated from the battlefield to Lacy House. Brinton remained at Lacy House for several more days, assisting in the surgery and collecting specimens.

71. Brinton, *PM*, 220.

72. Brinton, *PM*, 220.

73. On the history of scientific racism and exhumation involving the Army Medical Museum in the postwar years, see chapter 5, "The Unburied Dead," in Ann Fabian's *The Skull Collectors: Race, Science and America's Unburied Dead* (Chicago: University of Chicago Press, 2010), 165–204.

74. Walt Whitman, "My visits and distributions," holograph fragment, 1863, Berg Collection, New York Public Library. See also Kenneth M. Price, "Walt Whitman and Civil War Washington," *Leviathan* 16, no. 1 (2014): 121–34.

75. Whitman, *MDW*, 100.

76. Whitman, *MDW*, 101; Loving, *Walt Whitman: The Song of Himself*, 268.

77. Whitman, *MDW*, 16.

78. Artifacts displayed included the Declaration of Independence, Benjamin Franklin's printing press, Egyptian mummies, and a mosaic of Pompeii (Bradford, "Re-Collecting Soldiers," 135).

79. Henry Ellsworth, "Letter from Patent Office Commissioner Henry Ellsworth to Senator John Ruggles, Dec. 18, 1840," National Portrait Gallery, Smithsonian Institution, Washington, DC (http://www.npg.si.edu/exhibit/pob/).

80. Whitman, *MDW*, 16–17.

81. "United States South Seas Exploring Expedition (aka the Wilkes Expedition), 1838–1842," Harvard University Libraries Open Collections Program (http://ocp.hul.harvard.edu/expeditions/wilkes.html).

82. As Nathaniel Philbrick has established, "without the U.S. Exploring Expedition, there might never have been a national museum in Washington, D.C." In addition to the Smithsonian, "the U.S Botanic Garden, the U.S. Hydrographic Office, and the Naval Observatory all owe their existence, in varying degrees, to the expedition." Nathaniel Philbrick, "Young Ambition: Charles Wilkes' Antarctic Adventure," *MHQ: Quarterly Journal of Military History* 17 (2004): 84.

83. Whitman, *MDW*, 17.

84. Bradford has discussed the empathic function of anonymity as a vehicle for projection ("Re-Collecting Soldiers," 139).

85. Whitman, *MDW*, 16.

86. Whitman, *PW*, 1:95.

87. Whitman, *Corr.*, 1:112.

88. Whitman, "My visits and distributions."

89. Brinton, *PM*, 181–82.

90. Brinton, *PM*, 185–86.

91. Brinton, *PM*, 185–86.

92. Brinton, *PM*, 185.

93. Brinton, *PM*, 220.

94. Brinton, *PM*, 190.

95. Brinton, *PM*, 190–91.

96. Brinton, *PM*, 185–86.

97. Brinton, *PM*, 189.

98. John H. Brinton, "Closing Address, Army Medical School," *Journal of the American Medical Association* 26, no. 13 (March 28, 1896): 602. A subdued version of this incident is related in Brinton's memoir (*PM*, 190).

99. Brinton, *PM*, 190.

100. Mary Clemmer Ames, *Ten Years in Washington: Life and Scenes in the Nation's Capital, as a Woman Sees Them* (Hartford, CT: Worthington, 1873), 480.

101. Ames, *Ten Years in Washington*, 474.

102. Ames, *Ten Years in Washington*, 477.

103. Ames, *Ten Years in Washington*, 485.

104. Ames, *Ten Years in Washington*, 487.

105. Ames, *Ten Years in Washington*, 480.

106. Ames, *Ten Years in Washington*, 479.

107. Ames, *Ten Years in Washington*, 478.

108. Fabian, *Skull Collectors*, 5.

109. L. Bagger, "The Army Medical Museum in Washington," *Appleton's Journal* 9 (1873): 294.

110. Ames, *Ten Years in Washington,* 476. Brinton did not oversee the museum's relocation to Ford's Theatre. He was relieved of duty in the Surgeon General's Office in September 1864.

111. J. T. H. Connor and Michael Rhode, "Shooting Soldiers: Civil War Medical Images, Memory, and Identity in America," *Invisible Culture* 5 (2003). Web.

112. John Eric Erichsen, Lecture to University College Hospital, London, November 9, 1874, quoted by D. S. Lamb, *A History of the United States Army Medical Museum, 1862 to 1917* (Washington, DC: Government Printing Office, 1917), 210.

113. Ames, *Ten Years in Washington*, 475–76.

114. Kinney, "Making Capital," 185.

115. Walt Whitman, undated draft fragment, Harned Whitman Collection.

116. Whitman, undated draft fragment, Washington and Brooklyn Hospital Notebook (1864), Harned Whitman Collection.

117. Buinicki, *Walt Whitman's Reconstruction*, 54–55.

118. Whitman, *MDW*, 5. On Whitman's fascination with the impossibility of accurately recording the war, see Buinicki (*Walt Whitman's Reconstruction*, 6–8) and Thomas (*Lunar Light of Whitman's Poetry*, 234).

119. Whitman, *MDW*, 5–6.

120. Brinton, *PM*, 169–79.

121. Lenore Barbian, Paul S. Sledzik, and Jeffrey Reznick, "Remains of War: Walt Whitman, Civil War Soldiers, and the Legacy of Medical Collections," *Museum History Journal* 5, no. 1 (January 2012): 10.

122. Horace Traubel, *With Walt Whitman in Camden*, vol. 1 (Boston: Small Maynard, 1906), 178–79. Whitman probably wanted to include this material in *November Boughs* (1888), which included casualty statistics (Whitman, *PP*, 1233).

123. Faust, *Republic of Suffering*, 250–65.

124. Whitman, *MDW*, 57.

125. As Adam Bradford has observed, "Whitman sought to recover the bodies and preserve the identities of the Civil War's 'Million Dead' in the face of their material annihilation, much as he had done for his own body and identity in *Leaves of Grass*" (*Communities of Death*, 122).

126. Whitman, *SD*, 8; Whitman, *MDW*, 5.

127. Whitman, *Corr.*, 1:68.

128. Horace Traubel, *With Walt Whitman in Camden*, vol. 4, ed. Sculley Bradley (Philadelphia: University of Pennsylvania Press, 1953), 195.

129. Whitman, *MDW*, 5.

130. See chapter 9, "Calamus and the National Calamity," in Jerome Loving's *Walt Whitman: Song of Himself* (251–96).

131. Loving, *Walt Whitman: Song of Himself*, 268.

132. The wardmaster's room was the only room in a ward pavilion hospital that offered any semblance of privacy. Other small rooms may have included "a mess room, a scullery, a water closet, and an ablution room" (Roper, *Now the Drum of War*, 220).

133. William Stewart to Walt Whitman, July 17, 1865, *The Walt Whitman Archive*.

134. Alonzo S. Bush to Walt Whitman, December 22, 1863, *The Walt Whitman Archive*. See also Roper's analysis of this letter (*Now the Drum of War*, 218–19), and Jonathan Katz (*Love Stories: Sex between Men before Homosexuality* [Chicago: University of Chicago Press, 2001], 157–59).

135. Martin G. Murray, "Traveling with the Wounded: Walt Whitman and Washington's Civil War Hospitals," *The Walt Whitman Archive*, http://www.whitmanarchive.org/criticism/current/anc.00156.html.

136. Charles Shively, *Calamus Lovers: Walt Whitman's Working Class Camerados* (San Francisco: Gay Sunshine, 1987), 81–82.

137. Roper, *Now the Drum of War*, 220; Katz, *Love Stories*, 158.

138. Katz, *Love Stories*, 158; Roper, *Now the Drum of War*, 220.

139. Roper, *Now the Drum of War*, 220.

140. Eve Kosofsky Sedgwick, *Epistemology of the Closet* (Berkeley: University of California Press, 1990), 165. For example, as Price has observed, "Sexual passing is at the heart of the poem eventually titled 'Once I Pass'd through a Populous City' in which Whitman changed the pronouns from 'he' to 'she' to reorient (and disguise) his depiction of love, attachment, and loss" (Price, *To Walt Whitman, America*, 5).

141. Roper, *Now the Drum of War*, 221.

142. This is not to imply that Whitman's interest in or association with soldiers was entirely sexual. As numerous scholars have established, Whitman's primary motivation was compassion; his fundamental goal was to comfort the wounded and dying. If his treatment of soldiers had not been largely governed by discretion his presence would not have been accepted by physicians, nurses, Sanitary Commission officials, and the soldiers themselves (Roper, *Now the Drum of War*, 225).

143. Jonathan Katz discusses the common nineteenth-century custom of bed-sharing (*Love Stories*, 6). The act of taking a "bedfellow" was not necessarily sexual, though it certainly offered many erotic possibilities, which may or may not have been acted upon, given the circumstances. For specific Civil War examples, see Katz, *Love Stories*, 138–40.

144. Whitman, *SD*, 8.

145. Whitman, *MDW*, 57, 5–6.

146. Sedgwick, *Epistemology of the Closet*, 3.

147. Whitman, *MDW*, 5.

148. Whitman, *LG 1855*, 18.

149. Nicholas T. Rand, "Translator's Introduction," in *The Wolf Man's Magic Word: A Cryptonymy*, by Nicholas Abraham and Maria Torok (Minneapolis: University of Minnesota Press, 1986), lviii.

150. As Lawrence Johnson argues, the telling silences that orbit crypts eventually betray their existence. See "Cryptonymic Secretions," 117–31.

151. Rand, "Translator's Introduction," viii.

152. Nicholas Abraham and Maria Torok, *The Wolf Man's Magic Word: A Cryptonymy*, trans. Nicholas T. Rand (Minneapolis: University of Minnesota Press, 1986), 20.

153. Whitman, *MDW*, 48.

154. Abraham and Torok, *Wolf Man's Magic Word*, 20.

155. The term "magic word" originates from Freud's analysis of his patient Sergei Pankeiev, known as the "Wolf Man," who "created a secret magic word that, without betraying anybody, allowed him to achieve real or sublimated sexual gratification" (Abraham and Torok, *Wolf Man's Magic Word,* 40, 83). See also Lawrence Johnson, *The Wolf Man's Burden* (Ithaca, NY: Cornell University Press, 2001), 1–27.

156. Abraham and Torok, *Wolf Man's Magic Word*, 83.

157. Butler, *Antigone's Claim*, 80–88.

158. Whitman, *MDW*, 4.

159. See Lawrence Johnson, "Tracing Calculation (*Calque Calcul*) between Nicolas Abraham and Jacques Derrida," *Postmodern Culture* 10, no. 3 (2000): 2.

160. In the official report *Medical and Surgical History of the War of the Rebellion,* the surgeon general wrote: "Professor Leidy's autopsies constitute by far the most important contribution to the pathological anatomy of the non-ulcerative form of intestinal inflammation made during the war." Leonard Warren, *Joseph Leidy: The Last Man Who Knew Everything* (New Haven, CT: Yale University Press, 1998), 135.

161. *Catalogue of the Surgical Section of the US Army Medical Museum* (Washington, DC: Government Printing Office, 1866), 51.

162. Whitman, *MDW*, 3–4.

163. Warren, *Joseph Leidy*, 113.

164. Warren, *Joseph Leidy*, 220.

165. Undated draft lecture, Joseph Leidy Papers, College of Physicians of Philadelphia Historical Medical Library. I am grateful to Robert Hicks, director of the Mütter Museum & the Historical Medical Library, and to the scholarship of Christopher Willoughby for calling this document to my attention.

166. Marvin, "Body of the Text," 134.

167. Marvin, "Body of the Text," 142.

168. Undated draft letter, John H. Brinton Papers, College of Physicians of Philadelphia Historical Medical Library. See also W. S. Forbes, *History of the Anatomy Act of Pennsylvania* (Philadelphia Medical Publishing Company, 1898), 9.

169. Undated draft letter, John H. Brinton Papers, College of Physicians of Philadelphia Historical Medical Library.

170. See the Mütter Museum Catalog notes on "The Soap Lady": http://muttermuseum.org/exhibitions/the-soap-lady/.

171. William Hunt, M.D., *An Address Upon the Late Joseph Leidy, M.D.* (Philadelphia: Collins, 1892), 27–29. Delivered November 17, 1891, before alumni and students of the medical department of the University of Pennsylvania.

172. Humphrey, "Dissection and Discrimination," 822.

173. Richard A. Bauman, *Human Rights in Ancient Rome* (London: Routledge, 2000), 1–2. See also *The Oxford Encyclopedia of Ancient Greece and Rome*, vol. 1, eds. Michael Gagarin and Elaine Fantham (Oxford: Oxford University Press, 2010).

174. Bauman, *Human Rights in Ancient Rome*, 2.

175. Whitman, *LG 1867*, 107.

176. Bauman, *Human Rights in Ancient Rome*, 2.

177. Whitman, *MDW*, 57.

178. See Pauline Boss, *Ambiguous Loss: Learning to Live with Unresolved Grief* (Cambridge, MA: Harvard University Press, 1999).

179. Lenore Barbian, Paul Sledzik, and Jeffrey Reznick, former curators at the National Museum of Health and Medicine, recently identified four human specimens that originated from soldiers attended by Whitman. I am indebted to the groundbreaking discoveries of these curators. Significantly, the relationship between Whitman and these specific specimens in the Army Medical Museum has yet to be analyzed outside of a museum studies context. The authors confine their investigation to the soldiers' medical histories and Whitman's relationships with these four men; the correlations between Whitman and Brinton remain, until now, unexamined.

180. Whitman to Mrs. Irwin, May 1, 1865 (*Corr.*, 1:259).

181. Murray, "Traveling with the Wounded."

182. Memorandum by Willard Bliss, May 5, 1865, Otis Historical Archives, National Museum of Health and Medicine.

183. Oscar Cunningham's deterioration is narrated in Whitman's letters from May 6, 1864; May 10, 1864; May 25, 1864; June 3, 1864; and June 7, 1864. See this correspondence on *The Walt Whitman Archive.* On April 12, 1864, Whitman observed in his hospital notebook, "As I write this his leg is in a horrible condition, all livid & swollen out of shape—the chances are against him, poor fellow" (Harned Whitman Collection).

184. "Specimen No. 2254, Oscar Cunningham, Private, Co. J, 82nd Ohio Infantry," memorandum by Willard Bliss, May 5, 1865, Otis Historical Archives, National Museum of Health and Medicine.

185. Whitman to Louisa Van Velsor Whitman, June 3, 1864 (*Corr.*, 1:229).

186. "Specimen No. 2254."

187. "Specimen No. 2254."

188. "Specimen No. 1534, Oscar F. Wilbur, Private, Co. G, 154th New York Regiment," undated memorandum by Willard Bliss to John H. Brinton, Otis Historical Archives, National Museum of Health and Medicine.

189. "Specimen No. 1534."

190. Whitman misspells this soldier's last name, which is listed in hospital and museum records as Wilbur (*MDW*, 21).

191. Faust, *Republic of Suffering*, 7–10.

192. Whitman, *Corr.*, 1:127–29.

193. Whitman, *MDW*, 21.

194. "Notes on John Mahay, Surgical Specimen 2567," undated memorandum by John H. Brinton, Otis Historical Archives, National Museum of Health and Medicine.

195. Whitman, *MDW*, 21.

196. G. A. Otis and D. L. Huntington, *The Medical and Surgical History of the War of the Rebellion*, vol. 2, part 3 (Washington, DC: Government Printing Office, 1883), 294.

197. "Specimen No. 4077, Frank H. Irwin, Corporal, 93rd Pennsylvania Infantry," memorandum signed by D. W. Bliss, April 17, 1865, Otis Historical Archives, National Museum of Health and Medicine. Whitman spells this surname as Irvin. However, Martin Murray, a careful student of these soldiers, believes that Whitman may have gotten the name wrong, observing that certain military and pension records spell the name as "Irvin." See his work on annotating the soldiers of *Memoranda* at http://www.classroomelectric.org/volume2/price/memoranda/annotations/. National Museum of Health and Medicine records have transcribed the name as Irwin. However, ambiguity remains. The original archival documents that I inspected identify him once by initials in type ("Corporal F. H. I."), and three times in cursive handwriting (by three different authors) that could be read as either Irwin or Irvin. In their detailed account of the histories of these specimens, curators Barbian, Sledzik, and Reznick have identified the soldier as "Corporal Frank H. Irwin, 93rd Pennsylvania Infantry" (8–9).

198. As Barbian, Sledzik, and Reznick observed, the report is notably devoid of personal details compared to those written by Bliss, leading them to suspect that Munger was the author, although Bliss's signature appears on the report (22).

199. "Specimen No. 4077."

200. Whitman, *MDW*, 51.

201. In 2016, a volunteer working with the Civil War Widows' Pension Digitization Project at the National Archives uncovered a letter penned by Whitman on behalf of soldier Robert N. Jabo to his wife. For details on this document and two other recently discovered letters, see Kenneth M. Price and Jacqueline M. Budell, "Written by Walt Whitman, a Friend," *Prologue* 48, no. 2 (2016).

202. Whitman, *MDW*, 14.

203. Faust, *Republic of Suffering*, 9.

204. Faust, *Republic of Suffering*, 14.

205. Whitman, *Corr.*, 1:127–29.

206. Whitman, *LG 1855*, 55.

207. Brinton, *PM*, 186.

208. Mitchell, *Injuries of Nerves*, 348.

Chapter Three

1. For more on Whitman's relationship with Brown, see Katz, *Love Stories*, 152–58; Roper, *Now the Drum of War*, 254–55; Shively, *Calamus Lovers*, 69–88. See also Martin Murray, "Traveling with the Wounded," *The Walt Whitman Archive.*

2. Whitman, *NUPM*, 2:669.

3. Whitman, *NUPM*, 2:669. See also Whitman, *Corr.*, 1:259.

4. As Alfred Bollet writes, "The most prevalent misconception about Civil War medicine is that major surgery was usually performed without anesthesia." However, "patients were given an anesthetic only until they were insensitive to pain. At these superficial levels of anesthesia, patients remain in a state of excitement and delirium." Alfred Bollet, *Civil War Medicine: Challenges and Triumphs* (Tucson, AZ: Galen, 2002), 81. As Brown's case demonstrates, delivering a precise quantity was often a problem. "Many war-time amputations were done in the open air or at least in extremely well-ventilated rooms, as compared with the closed operating rooms or theatres of civilian practice . . . . The patient was under the influence of anesthetic for a relatively short time, and because he was outside or in a well-ventilated room, the anesthetic was easily expelled from his system." Michael A. Franchetti, "Trauma Surgery during the Civil War," *Southern Medical Journal* 86, no. 5 (1993): 553.

5. Whitman, *NUPM*, 2:669.

6. Ambrose Paré, *The Works of That Famous Chirurgeon Ambrose Parey*, trans. T. Johnson (London: M. Clark, 1649), 773.

7. This lacuna in medical history is believed to be the result of amputees or their doctors not reporting the phenomenon because such symptoms "would [have been] tantamount to losing one's reason and/or admitting that the devil or some other supernatural forces had gained entrance into the body. This would, because of the status of medicine and society in general prior to the nineteenth century, leave one's self wide open to all kinds of punishments." Julius Hoffman, "Phantom Limb Syndrome: A Critical Review of Literature," *Journal of Nervous and Mental Disease* 119 (1954): 261. See also Cassandra S. Crawford, *Phantom Limb: Amputation, Embodiment and Prosthetic Technology* (New York: New York University Press, 2014), 108–12.

8. Hacker, "Census-Based Count," 310; Cervetti, *S. Weir Mitchell*, 82; Faust, *Republic of Suffering*, xi.

9. Laurann Figg and Jane Farrell-Beck, "Amputation in the Civil War: Physical and Social Dimensions," *Journal of the History of Medicine and Allied Sciences* 48 (1993): 454.

10. Robert I. Goler, "Loss and the Persistence of Memory: 'The Case of George Dedlow' and Disabled Civil War Veterans," *Literature and Medicine* 23, no. 1 (Spring 2004): 163.

11. Whitman, *MDW*, 46.

12. Goler, "Loss and the Persistence of Memory," 163; Cervetti, *S. Weir Mitchell*, 82.

13. Walt Whitman to Margaret S. Curtis, October 4, 1863, *The Walt Whitman Archive*.

14. Whitman, *MDW*, 12.

15. Mitchell, *Injuries of Nerves*, 348.

16. Whitman, *LG 1891*, 243.

17. Military casualties in Iraq and Afghanistan are impacted by two major differences between these wars and previous conflicts: "(1) Many veterans survive injuries that would have killed them in past wars, and (2) improvised explosive device attacks have caused 'polytraumatic' injuries (multiple amputations; brain injury; severe facial trauma or blindness). War-related military amputations now occur at double the rate seen in previous wars." James Geiling, Joseph M. Rosen, and Ryan D. Edwards, "Medical Costs of War in 2035: Long-Term Care Challenges for Veterans of Iraq and Afghanistan," *Military Medicine* 177, no. 11 (2012): 1235–38. Multiple hypotheses have been put forward to explain the higher amputation rates in Iraq and Afghanistan. Soldiers attacked with automatic weapons are often struck by multiple gunshots. While bullets striking the chest or abdomen were more likely to have been fatal during earlier conflicts, those impacts are much more survivable to current soldiers who wear modern body armor. However, the arms and legs remain unprotected by body armor, so a larger percentage of surviving soldiers now have to contend with the loss of limbs. Factors that raise the overall survival rate may indirectly increase the number of amputees. Battle wounds are often infected by dirt, shreds of the soldier's clothing, and other debris, but advances have been made in the treatment of such infections. New blood-clotting factors have reduced the risk of death from blood loss. Simply put, medical advancements mean more soldiers *survive* amputation and its aftermath. Matthew S. Goldbert, "Updated Death and Injury Rates of U.S. Military Personnel during the Conflicts in Iraq and Afghanistan," Congressional Budget Office Working Paper (2014): 11–12. See also H. N. Lukes, "Causalgia of the Heart: Phantom Limb Syndrome, Disability and Queer Feeling," *Women & Performance* 19, no. 2 (2009): 240.

18. V. S Ramachandran and Sandra Blakeslee discovered that the use of a mirror box to replicate the intact limb significantly alleviated phantom pain. The majority of patients studied experienced temporary relief, and many eventually lost their phantom through continued mirror box therapy. Ramachandran and Blakeslee, *Phantoms in the Brain: Probing the Mysteries of the Human Mind* (New York: William Morrow, 1999), 49. See also Crawford, *Phantom Limb*, 187–88.

19. Peter W. Halligan, "Phantom Limbs: The Body in Mind," *Cognitive Neuropsychiatry* 7, no. 3 (2002): 25.

20. V. S. Ramachandran and W. Hirstein, "The Perception of Phantom Limbs," *Brain* 121 (1998): 1603–30.

21. Halligan, "Phantom Limbs," 261.

22. Crawford, *Phantom Limb*, 112–13.

23. Halligan, "Phantom Limbs," 261.

24. Crawford, *Phantom Limb*, 33.

25. Silas Weir Mitchell, "Phantom Limbs," *Lippincott's Magazine of Popular Literature & Science* 8 (1871): 567. Jonah Leher's chapter, "Walt Whitman: The Substance of Feeling," draws intriguing connections between Weir Mitchell and Whitman, particularly in relation to "I Sing the Body Electric." *Proust Was a Neuroscientist* (New York: Houghton, 2007), 1–24.

26. Whitman, *MDW*, 15.

27. Walt Whitman, hospital notebook, 1863, Harned Whitman Collection. This notebook also contains brief entries on two other amputees: "Chris Miller, bed 19, Ward I, Armory Hospital, left leg amputated—father living—mother dead—no relatives or friends here to see him—wants some apples—wants tobacco. . . . Sam Elliott, father John Elliot, from Cumberland Valley . . . Cavalry cleaning his pistol shot himself—good family—May 5, operated on, took chloroform, took his leg off—he died after the operation." See also *NUPM*, 2:606.

28. Caruth, *Unclaimed Experience,* 8.

29. *NUPM*, 2:878.

30. Whitman, *MDW*, 48.

31. Whitman, *MDW*, 27.

32. Reynolds, *Walt Whitman's America*, 259–60.

33. McGarry, *Ghosts of Futures Past*, 130

34. Reynolds, *Walt Whitman's America*, 259–260. See also Adam Crabtree, *From Mesmer to Freud: Magnetic Sleep and the Roots of Psychological Healing* (New Haven, CT: Yale University Press, 1993).

35. The phrase *romance of surgery* appears in an 1862–63 notebook underneath the notes on the "case of boy in Ward 12." Harned Whitman Collection.

36. Whitman, *MDW*, 30

37. Whitman, *MDW*, 27.

38. Whitman to Moses Lane, March 13, 1864, *The Walt Whitman Archive.*

39. Whitman, *MDW*, 28.

40. Whitman, *MDW*, 27.

41. Mitchell, "Phantom Limbs," 563.

42. Mitchell, "Phantom Limbs," 565.

43. Silas Weir Mitchell Papers Finding Aid, College of Physicians of Philadelphia Historical Medical Library.

44. Several of Mitchell's medical cases reappeared in his fiction. Cervetti observes: "Instances of intertextuality are numerous where passages repeat themselves and reappear . . . both the medical and literary writing deal extensively with causalgia and phantom limbs, and some of the cases are the same. The case of Joseph H. Corliss, age twenty-seven, appears three times in Mitchell's writing—in *Gunshot Wounds*, in *Injuries of Nerves*, and in 'The Case of George Dedlow' (as the hysterical Dane)" (Cervetti, *S. Weir Mitchell*, 80).

45. Although this treatment evolved from his work with Civil War soldiers suffering the effects of exhaustion and traumatic stress, the "Rest Cure" became gendered as a treatment for hysteria. Coupled with Mitchell's vocal sexism, it left a dark

stain on his medical legacy. Mitchell's reputation never recovered from his (mis)treatment of Charlotte Perkins Gilman, who immortalized the "Rest Cure" as tantamount to violence and named the neurologist himself as an assailant in her gothic tale of an imprisoned wife's descent into madness, "The Yellow Wall-Paper." In her autobiography, Gilman writes that "The Yellow Wall-Paper" describes a mental "breakdown," "beginning something as mine did, and treated as Dr. S. Weir Mitchell treated me with what I considered the inevitable result, progressive insanity." Charlotte Perkins Gilman, *The Living of Charlotte Perkins Gilman: An Autobiography* (1935; Madison: University of Wisconsin Press, 1963), 119. As David Schuster argues, "over the past thirty-five years Mitchell has developed a notorious reputation as a misogynist who sought to infantilize his female patients . . . due largely to his failed treatment of the feminist writer and intellectual Charlotte Perkins Gilman." David Schuster, "Personalizing Illness and Modernity: S. Weir Mitchell, Literary Women, and Neurasthenia, 1870–1914," *Bulletin of the History of Medicine* 79, no. 4, (2005): 718. Yet, while Gilman is certainly the most famous of Mitchell's maligned hysterics, his reputation is, as Molly McGarry establishes, a matter of published record and is based on his paternalistic, at times violent, treatment of women (*Ghosts of Futures Past*, 147–48). In one instance he claimed to have "cured" a woman believed to be terminally ill by setting fire to her bed, after which he calmly remarked to onlookers, "she will run out of the door in two minutes; I set her sheets on fire. A case of hysteria." Quoted in J. M. S. Pearce, "Silas Weir Mitchell and the 'Rest Cure,'" *Journal of Neurology, Neurosurgery and Psychiatry* 75, no. 381 (2004): 42. On connections between Gilman and Whitman, see Joann P. Krieg, "Charlotte Perkins Gilman and the Whitman Connection," *Walt Whitman Quarterly Review* 1 (1984): 21–25.

46. Kaplan, *Walt Whitman: A Life*, 296.

47. Finch, "Aging, Inflammation, and the Body Electric," 68.

48. Walt Whitman to Thomas Jefferson Whitman, March 6, 1863, *The Walt Whitman Archive.*

49. Kaplan, *Walt Whitman: A Life*, 298.

50. Loving, *Walt Whitman: Song of Himself*, 479; Finch, "Aging, Inflammation, and the Body Electric," 68.

51. Whitman, *Corr.*, 1:236.

52. Walt Whitman to Thomas Jefferson Whitman, March 6, 1863, *The Walt Whitman Archive.*

53. Loving, *Walt Whitman: Song of Himself*, 308.

54. Whitman suffered an earlier "sunstroke" in 1858, which Reynolds attributes to high blood pressure (Reynolds, *Walt Whitman's America*, 375; Turpin, introduction, 150).

55. In his daybook entry for April 18, 1878, Whitman recorded his initial consultation with the renowned neurologist (Charles E. Feinberg Collection, Library of Congress). Between 1873 and 1888 he suffered six strokes, which eventually left him predominantly wheelchair-bound. A series of paralytic strokes beginning June 2, 1888, nearly proved fatal; he was virtually bedridden for almost a year (Loving, *Walt Whitman: Song of Himself*, 460–61).

56. Whitman to Louisa Orr Whitman, April 13–14, 1878, *The Walt Whitman Archive.* According to this letter, Mitchell withheld his diagnosis on this first visit: "He did not

express any opinion particular—said he would tell me next time—examined my heart by auscultation—said there was nothing at all the matter with *that*. . ."

57. Walt Whitman to Louisa Orr Whitman, April 13–14, 1878, footnote 4, *The Walt Whitman Archive.*

58. Whitman's Commonplace Book, Charles E. Feinberg Collection of the Papers of Walt Whitman, 1839–1919, Library of Congress. See also Walt Whitman to Louisa Orr Whitman, footnote 4, *The Walt Whitman Archive.*

59. Subsequent correspondence indicates that Whitman began to improve after his visits with Mitchell. See Walt Whitman to George W. Waters, April 23, 1878, footnote 2, *The Walt Whitman Archive.* On Mitchell's prescription of sunbaths, see Harold Aspiz, "Specimen Days: The Therapeutics of Sun-Bathing," *Walt Whitman Quarterly Review* 1 (1983): 48–50. Aspiz establishes that Whitman already practiced this treatment prior to his consultation with Mitchell.

60. Ernest Earnest, *S. Weir Mitchell: Novelist and Physician* (Philadelphia: University of Pennsylvania Press, 1950), 115.

61. For details on Mitchell's indignation following the publication of Traubel's first two volumes, see Cervetti, *S. Weir Mitchell*, 171–73. The following comment, from Traubel's second volume, is indicative of the overall tone: "I can't say that he's a world-author—he don't hit me for that size—but he's a world-doctor for sure—leastwise everybody says so and I join in" (Traubel, *With Walt Whitman in Camden*, vol. 2 [New York: D. Appleton, 1908], 268). After Mitchell's death, other revelatory remarks on Whitman's impressions of Mitchell were published in the subsequent volumes of *With Walt Whitman in Camden.* For example, Whitman tells Traubel, "Mitchell of late years has been bitten with the desire to compose, compose—that curse of curses: has written volumes: very bad, too—awful in their inadequacy: but personally he is a man to meet, to know" (Traubel, *With Walt Whitman in Camden*, 3:339).

62. Peter B. Martens, "War, Whitman and Osler," *Literature and Medicine* 16, no. 2 (1997): 215.

63. Cervetti, *S. Weir Mitchell*, 171.

64. Cervetti, *S. Weir Mitchell*, 171; Jennifer A. Hynes, "Silas Weir Mitchell (1829–1914)," in *Walt Whitman: An Encyclopedia,* eds. J. R. LeMaster and Donald D. Kummings (New York: Garland, 1998). Available on *The Walt Whitman Archive.*

65. Traubel, *With Walt Whitman in Camden*, 6:107.

66. Traubel, *With Walt Whitman in Camden*, 6:107.

67. For example, "On April 15, 1886, Mitchell anonymously donated $200 along with his purchase of tickets for Whitman's annual Lincoln lecture at the opera house on Chestnut Street. That same year, the Philadelphia literati held a lecture and birthday party for Whitman at Reisser's restaurant. John and Silas Weir both attended, and the elder Mitchell was one of the speakers" (Cervetti, *S. Weir Mitchell*, 171).

68. Cervetti, *S. Weir Mitchell*, 48.

69. Silas Weir Mitchell, George R. Morehouse, and William W. Keen, *Gunshot Wounds and Other Injuries of the Nerves* (Philadelphia: Lippincott, 1864), 23.

70. Mitchell, Morehouse, and Keen, *Gunshot Wounds,* 24.

71. Walt Whitman, "A Sight in Camp in the Day-Break Grey and Dim," *LG 1867*, 46a.

72. Mitchell, Morehouse, and Keen, *Gunshot Wounds*, 27.

73. Mitchell, Morehouse, and Keen, *Gunshot Wounds*, 27.

74. Marc-Antoine Crocq and Louis Crocq, "From Shell Shock and War Neurosis to Posttraumatic Stress Disorder: A History of Psychotraumatology," *Dialogues in Clinical Neuroscience* 2, no. 1 (2000): 50.

75. Charles E. Feinberg, "Walt Whitman and His Doctors," *Archives of Internal Medicine* 114 (December 1964): 836.

76. Whitman to Peter Doyle, September 5, 1873, quoted in Feinberg, "Walt Whitman and His Doctors," 837.

77. Feinberg, "Walt Whitman and His Doctors," 837.

78. Feinberg, "Walt Whitman and His Doctors," 835.

79. Mitchell, Morehouse, and Keen, *Gunshot Wounds*, 26.

80. Mitchell, "Phantom Limbs," 565.

81. Halligan, "Phantom Limbs," 251; J. Herman, "Phantom Limb: From Medical Knowledge to Folk Wisdom and Back," *Annals of Internal Medicine* 128 (1998): 76–78.

82. In October 1892 Weir Mitchell wrote to Charles Keyson, an amputee soldier he had attended at Turner's Lane, posing questions that were replicated in a standardized questionnaire circulated by Weir Mitchell and his son John. Silas Weir Mitchell Papers, College of Physicians of Philadelphia Historical Medical Library. Hereafter "Mitchell Papers, CPP."

83. A letter to John K. Mitchell from Millings (illeg.), on Army Medical Museum stationery (September 21, 1892) reports, "your placards have received a good deal of attention," and includes the names and addresses of amputees who left their details with the museum staff. Mitchell Papers, CPP.

84. John K. Mitchell, form letter attached to blank surveys, September 4, 1893, Mitchell Papers, CPP.

85. Silas Weir Mitchell, "The Case of George Dedlow," *Atlantic Monthly* 18, no. 105 (July 1866): 1–11. Reprinted in *The Autobiography of a Quack and The Case of George Dedlow* (New York: Century, 1900), 149.

86. Mitchell, "Case of George Dedlow," 129.

87. Mitchell, "Case of George Dedlow," 133.

88. He claimed to have written the story in response to a friend's virtually identical query: "How much of a man would have to be lost in order that he should lose any portion of his sense of individuality?" Robert I. Goler, "Loss and the Persistence of Memory: 'The Case of George Dedlow' and Disabled Civil War Veterans," *Literature and Medicine* 23, no. 1 (2004): 171.

89. Mitchell, "Case of George Dedlow," 149.

90. Mitchell, "Case of George Dedlow," 149.

91. Mitchell, "Case of George Dedlow," 139.

92. Mitchell, "Case of George Dedlow," 116.

93. Mitchell Papers, CPP.

94. Mitchell Papers, CPP.

95. The catalog numbers of Dedlow's legs did not match any actual specimens from the Army Medical Museum. Silas Weir Mitchell, "The Medical Department in the Civil War," *Journal of the American Medical Association* 62 (1914): 1448.

96. Mitchell, "Case of George Dedlow," 148.

97. Mitchell, "Case of George Dedlow," 149.

98. Brinton, *PM*, 186; Mitchell, "Phantom Limbs," 567. Dedlow's desire to "rejoin" his "corporeal family" in the afterlife foreshadows an ongoing refrain in neurological research. As Brinton observed at the Army Medical Museum, many amputees remained fixated on their limbs' "final resting place," an anxiety that medical science has not always viewed compassionately. Concern for the limbs' afterlives may manifest in the desire for burial, cremation, freezing, taxidermy, or wet specimen preservation, and often originates from the "hope of a reunion with the severed limb." In 1952, Lawrence Kolb described the tendency of some patients to "panic about whether the amputated hand will be handled with respect and tenderness by the surgical team. . . they may secretly weep or become unduly anxious." Echoing Brinton's condescension toward the soldier who sought to repossess his limb, Kolb dismissed amputees' concerns for their detached parts as psychologically disordered: "it is the highly neurotic persons who have a need for any complicated ritualistic burial." Contemporary neurology is far more empathic in theorizing amputees' relationships to their disembodied limbs. Crawford describes the anxiety experienced by many at the thought of their severed parts as medical waste or anatomical material: "not knowing whether the limb . . . found its way to an incinerator, the trash heap out back, or the resident laboratory in the name of medical inspection and experimentation was for some amputees utterly unbearable" (Crawford, *Phantom Limb*, 136).

99. Brinton, *PM*, 602.

100. Brinton, *PM*, 186.

101. Mitchell, "The Medical Department in the Civil War," 1448.

102. He confessed that "the author" took "advantage of the freedom accorded to a writer of fiction," never imagining that his "humorous sketch, with its absurd conclusion, would for a moment mislead any one." Mitchell, "Phantom Limbs," 564.

103. Cervetti, Mitchell's most recent and comprehensive biographer, describes the story's spiritualist ending as absurdist satire, citing his cynical response to a séance attended many years later with William James, which he described as a "babble of utter nonsense" (*S. Weir Mitchell*, 78). Mitchell also claimed that the story was sent to the *Atlantic Monthly* without his knowledge (Goler, "Loss and the Persistence of Memory," 172).

104. Mitchell, "Phantom Limbs," 566.

105. Mitchell, "Phantom Limbs," 565–66.

106. Mitchell, "Phantom Limbs," 566.

107. Mitchell, "Phantom Limbs," 567.

108. As Crawford establishes, researchers have found that "phantoms often maintained the fixed, relaxed, flexed, or twisted posture of preamputated limbs, particularly the position of the limb *just prior* to loss" (*Phantom Limb*, 36).

109. Mitchell, "Phantom Limbs," 568.

110. Ronald Melzack, "Phantom Limbs," *Scientific American* 266 (1992): 120–26. See also Halligan, "Phantom Limbs," 254.

111. Mitchell, "Phantom Limbs," 566.

112. Mitchell, "Phantom Limbs," 566–67.

113. Mitchell, "Phantom Limbs," 566.

114. Michael Moon, *Disseminating Whitman: Revision and Corporeality in Leaves of Grass* (Cambridge, MA: Harvard University Press, 1991), 174.

115. Halligan, "Phantom Limbs," 252.

116. Whitman, *MDW*, 48.

117. Roper, *Now the Drum of War*, 45.

118. Foucault, *History of Sexuality*, vol. 1, p. 43, emphasis mine.

119. Whitman, *LG 1891*, 243–44.

120. Whitman, *MDW*, 54.

121. Whitman, *LG 1891*, 46.

122. Sedgwick, *Epistemology of the Closet*, 236.

123. Whitman, *MDW*, 48.

124. John Traill Taylor, *The Veil Lifted: Modern Developments of Spirit Photography* (London: Whitaker, 1894), vi.

125. McGarry, *Ghosts of Futures Past*, 14.

126. Reynolds, *Walt Whitman's America*, 263.

127. Whitman, *LG 1891*, 100. As Huck Gutman writes, "Whitman always sought the comradeship he celebrated so powerfully in the 'Calamus' sequence. The Civil War, despite or perhaps because of its violence, disruption, and widespread suffering, paradoxically allowed him to experience that comradeship on the most profound level." Huck Gutman, "Drum-Taps," in *Walt Whitman: An Encyclopedia,* eds. J. R. LeMaster and Donald D. Kummings (New York: Garland, 1998), sourced from *The Walt Whitman Archive.*

128. Hans Fluck and Rita Jaspersen-Schib, *Medicinal Plants and Their Uses*, trans. J. M. Rowson (London: W. Foulsham, 1976), 32.

129. Quoted by Gary Schmidgall in *Walt Whitman: A Gay Life* (New York: Dutton, 1997), 70.

130. Farland, "Decomposing City," 817.

131. Whitman, *LG 1891*, 243.

132. Lukes, "Causalgia of the Heart," 240.

133. Mitchell, Morehouse, and Keen, *Gunshot Wounds*, 10.

134. Quoted in Jennifer Travis, *Wounded Hearts: Masculinity, Law and Literature in American Culture* (Chapel Hill: University Press of North Carolina, 2005), 23.

135. Whitman, *NUPM*, 1:125.

136. Freud, "MM," 262.

137. Whitman, *Corr.*, 1:142.

138. Freud, "MM," 262.

139. Freud, "MM," 244.

140. Whitman, *MDW*, 4.

141. Sedgwick, *Epistemology of the Closet*, 236.

142. Whitman, *LG 1891*, 452.

143. Eric T. Dean, *Shook Over Hell: Post-Traumatic Stress, Vietnam, and the Civil War* (Cambridge, MA: Harvard University Press, 1997), 115–16. See also Svetlana Boym, chapter 1, "From Cured Soldiers to Incurable Romantics," in *The Future of Nostalgia* (New York: Basic Books, 2001), 3–19.

144. Hofer quoted in Fred Davis, *Yearning for Yesterday: A Sociology of Nostalgia* (New York: Free, 1979), 2. See also David Lowenthal, "Past Time, Present Place: Landscape and Memory," *Geographical Review* 65, no. 1 (1975): 1.

145. Mitchell, "Phantom Limbs," 565.

146. Freud, *Studies on Hysteria*, 221.

147. Hofer quoted in Jean Starobinski, "The Idea of Nostalgia," *Diogenes* 54 (1966): 87.

148. Hofer theorized that nostalgia "admits no remedy other than a return to the homeland." George Rosen, "Nostalgia: A 'Forgotten' Psychological Disorder," *Psychological Medicine* 5 (1975): 344.

149. Hofer quoted in Rosen, "Nostalgia," 343.

150. Starobinski, "Idea of Nostalgia," 94.

151. Hofer quoted in Starobinski, "Idea of Nostalgia," 97.

152. Whitman to Lewis K. Brown, Brooklyn, July 11, 1864, Berg Collection, New York Public Library.

153. Rosen, "Nostalgia," 351.

154. Roberts Bartholow, *Manual of Instructions for Enlisting and Discharging Soldiers* (Philadelphia: Lippincott, 1863), 21.

155. On symptoms of Civil War era nostalgia, see Donald Lee Anderson and Godfrey Tryggve, "Nostalgia and Malingering in the Military during the Civil War," *Perspectives in Biology and Medicine* 28 (1984). See also Dean, *Shook Over Hell*, 129; and Rosen, "Nostalgia," 348.

156. Anderson and Tryggve, "Nostalgia and Malingering," 157.

157. Anderson and Tryggve, "Nostalgia and Malingering," 157–58.

158. Mitchell, "The Medical Department in the Civil War," 1449.

159. Anna O (Bertha Pappenheim) famously described her therapy with Josef Breuer as a "talking cure." See John Launer, "Anna O and the 'Talking Cure,'" *QJM* 98, no. 6 (2005): 465–66. Although "Freud described his own therapy as a modification of Mitchell's 'rest cure' and had favorably reviewed many of his writings, Mitchell himself mocked mind-oriented approaches like Freud's and purportedly threw a psychoanalytic text into the fire, demanding, 'Where did this filthy thing come from'" (McGarry, *Ghosts of Futures Past*, 148).

160. H. S. Huidekoper to Silas Weir Mitchell, February 10, 1906, Mitchell Papers, CPP.

161. W. S. Bigelow to Silas Weir Mitchell, March 8, 1911, Mitchell Papers, CPP.

162. Mitchell also corresponded with the Shakespeare scholar Horace Howard Furness, the author and critic William Dean Howells (editor at the *Atlantic* at the time of Dedlow's publication), and the poet James Whitcomb Riley on this topic. Mitchell Papers, CPP.

163. Sigmund Freud, *The Interpretation of Dreams*, trans. A. A. Brill (1913; New York: Dover, 2015), 93.

164. Crawford, *Phantom Limb*, 135. See also Peter Brugger, "The Phantom Limb in Dreams," *Conscious Cognition* 17, no. 4 (2008): 1272–80; T. Mulder, J. Hochstenbach, P. U. Dijkstra, and J. H. Geertzen, "Born to Adapt, but Not in Your Dreams," *Conscious Cognition* 17, no. 4 (2008): 1266–71.

165. Whitman, *LG 1891*, 367.

166. Loving, *Walt Whitman: Song of Himself*, 479; Kaplan, *Walt Whitman: A Life*, 296; Thomas, *Lunar Light of Whitman's Poetry*, 278.

167. Kaplan, *Walt Whitman: A Life*, 296.

168. Walt Whitman, autograph letter signed to Lewis K. Brown, Brooklyn, July 11, 1864, Berg Collection, New York Public Library.

169. Anderson and Tryggve, "Nostalgia and Malingering," 157.

170. Kaplan, *Walt Whitman: A Life*, 295–96.

171. Traubel, *With Walt Whitman in Camden*, 2:137.

172. David Hsu, "Walt Whitman: An American Civil War Nurse Who Witnessed the Advent of Modern American Medicine," *Archives of Environmental & Occupational Health* 65, no. 4 (2010): 238. Several symptoms of nostalgia preceded Whitman's collapse, including "hectic fever," "a dull pain in the head," "throbbing of the temporal arteries," "anxious expression of the face," "watchfulness," and "a general wasting of all the vital powers" (Anderson and Tryggve, "Nostalgia and Malingering," 157).

173. Hsu, "Walt Whitman: An American Civil War Nurse," 238.

174. Paul Ricoeur, *Memory, History, Forgetting*, trans. K. Blamey and D. Pellauer (Chicago: University of Chicago Press, 2004), 69.

175. Traubel, *With Walt Whitman in Camden*, vol. 7 (Carbondale: Southern Illinois University Press, 1992), 356.

176. Whitman, *LG 1891*, 242.

177. Whitman, *MDW*, 3.

178. Jane Goodall and Christopher Lee, eds., *Trauma and Public Memory* (New York: Palgrave Macmillan Memory Studies Series, 2015), 14. On temporality and mourning in Whitman's milieu, see Dana Luciano, *Arranging Grief: Sacred Time and the Body in Nineteenth-Century America* (New York: New York University Press, 2007).

179. Goodall and Lee, *Trauma and Public Memory*, 14.

180. The opening stanza of this poem recalls these lines from a draft poem fragment in Whitman's December 1862 Fredericksburg notebook, analyzed in full in the first chapter: "How the dead lie / Some lie on their back with faces up and arms extended." Harned Whitman Collection.

181. Whitman, *LG 1867*, 180.

182. Whitman, *LG 1891*, 411–12.

183. Bradford, "Re-Collecting Soldiers," 101.

184. Buinicki, *Walt Whitman's Reconstruction*, 8.

185. Whitman, *PP*, 1214.

186. Derrida, *Aporias*, 61–62.

187. Mitchell, "Case of George Dedlow," 149.

Chapter Four

1. Faust, *Republic of Suffering*, 70–74.

2. Faust, *Republic of Suffering*, 66–70.

3. One of the most famous war photographs, *Harvest of Death*, shot by Timothy O'Sullivan, was so titled because it captured the Gettysburg battlefield (a farm field) covered with corpses. See Ed Folsom, "Nineteenth-Century Visual Culture," in *A Companion to Walt Whitman*, ed. Donald D. Kummings (Oxford: Blackwell, 2009), 281.

4. Cavitch, *American Elegy*, 239.

5. Walt posed for a photograph in Brady's studio sometime in 1862. For more on Whitman's relationship to photography and his connection with Gardner, see Folsom,

"Nineteenth-Century Visual Culture," 282. See also Daniel Mark Epstein, *Lincoln and Whitman: Parallel Lives in Civil War Washington* (New York: Random House, 2004), 176–77.

6. Faust, *Republic of Suffering*, 66.

7. William A. Frassanito, *Antietam: The Photographic Legacy of America's Bloodiest Day* (New York: Charles Scribner's Sons, 1978), 17.

8. Quoted in Keith Davis, "'A Terrible Distinctness': Photography of the Civil War Era," in *Photography in Nineteenth-Century America*, ed. Martha A. Sandweiss (New York: Harry N. Abrams, 1991), 130–79. See also Bradford, "Re-Collecting Soldiers," 137.

9. Keith Davis, "A Terrible Distinctness," 150.

10. Keith Davis, "A Terrible Distinctness," 150.

11. Whitman, *PP*, 1214; Whitman, *MDW*, 9–10.

12. Keith Davis, "A Terrible Distinctness," 150.

13. Timothy Sweet has argued that Civil War photographers often imitated the work of nineteenth-century landscape painters by evoking pastoral images, even in the aftermath of massacres (*Traces of War*, 109).

14. Schantz, *Awaiting the Heavenly Country*, 184–87. See also Laderman, *Sacred Remains*, 148–51; Sweet, *Traces of War*, 107–10.

15. Corpses were often rearranged to create more arresting images, and sometimes living soldiers were coaxed by photographers into posing as corpses (Cavitch, *American Elegy*, 239).

16. Schantz, *Awaiting the Heavenly Country*, 186–87.

17. Laderman, *Sacred Remains*, 149.

18. Oliver Wendell Holmes, "Doings of the Sunbeam," in *Soundings, from the "Atlantic"* (Boston: Ticknor and Fields, 1866), 266–67.

19. Roland Barthes, *Camera Lucida: Reflections on Photography,* trans. Richard Howard (New York: Hill and Wang, 1982).

20. Whitman, *MDW*, 17.

21. Whitman, *MDW*, 14.

22. Whitman, *MDW*, 14–15.

23. Faust, *Republic of Suffering*, 102; Whitman, *MDW*, 57.

24. Whitman, *MDW*, 57.

25. See Folsom, "Walt Whitman and the Civil War," 18–19.

26. See Faust, *Republic of Suffering*, 250–65. Cavitch has argued that Whitman's war texts subvert nationalistic and statistical mechanisms for calculating the costs of war, participating instead in a "writing of 'remains' that is itself characterized by patchwork, discontinuity, and open-endedness" (Cavitch, *American Elegy*, 238).

27. Faust, *Republic of Suffering*, 250.

28. Walt Whitman, "A Voice from Death," The Johnstown, Penn., cataclysm, May 31, 1889. Available on *The Walt Whitman Archive.*

29. Whitman, "A Voice from Death."

30. Whitman, "A Voice from Death."

31. Faust, *Republic of Suffering*, 57.

32. Roper, *Now the Drum of War*, 278.

33. Faust, *Republic of Suffering*, 58.

34. Faust, *Republic of Suffering*, 57.

35. Quoted in Morris, *Better Angel*, 91.

36. Whitman, *MDW*, 5–6.

37. Whitman, "Ashes of Soldiers," in *LG 1891*, 411–12.

38. Technically, this was not a separate edition of *Leaves of Grass,* but a reprinting of the 1881–82 edition supplemented by annexes (see *The Walt Whitman Archive*). "A Twilight Song" was included in the "Good-Bye My Fancy" annex.

39. Whitman, *LG 1891*, 460–61.

40. Whitman, *LG 1891*, 460–61.

41. To choose but one example, from the first edition: "And as to you corpse I think you are good manure, but that does not offend me, / I smell the white roses sweet-scented and growing, / I reach to the leafy lips" (Whitman, *LG 1855*, 54).

42. Laderman, *Sacred Remains*, 116.

43. Laderman, *Sacred Remains*, 113.

44. Laderman, *Sacred Remains*, 113.

45. Faust, *Republic of Suffering*, 94.

46. Laderman, *Sacred Remains*, 115.

47. Faust, *Republic of Suffering*, 95.

48. Faust, *Republic of Suffering*, 95.

49. Faust, *Republic of Suffering*, 94.

50. Faust, *Republic of Suffering*, 212.

51. David W. Blight, "Decoration Days: The Origins of Memorial Day in the North and South," in *The Memory of the Civil War in American Culture*, eds. Alice Fahs and Joan Waugh (Chapel Hill: University of North Carolina Press, 2004), 97; Faust, *Republic of Suffering*, 213; Blight, "Decoration Days," 97.

52. Faust, *Republic of Suffering*, 223.

53. Identification of the Confederate dead was slower and more problematic, due to insufficient resources (Blight, "Decoration Days," 97).

54. Special Order no. 132 in "Report of Captain J. M. Moore," in *Executive Documents Printed by Order of the House of Representatives during the First Session of the Thirty-Ninth Congress, 1865–66* (Washington, DC: Government Printing Office, 1866). Quoted in Faust, *Republic of Suffering*, 214.

55. Faust, *Republic of Suffering*, 215.

56. Faust, *Republic of Suffering*, 215.

57. Faust, *Republic of Suffering*, 215.

58. General Montgomery C. Meigs, quoted by Faust, *Republic of Suffering*, 219.

59. Faust, *Republic of Suffering*, 220.

60. Faust, *Republic of Suffering*, 220.

61. Faust, *Republic of Suffering*, 220.

62. Faust, *Republic of Suffering*, 220.

63. Faust, *Republic of Suffering*, 222.

64. Whitman, *MDW*, 57.

65. Caruth, *Unclaimed Experience,* 4.

66. Whitman, *LG 1867*, 43a.

67. Whitman, *NUPM*, 2:493. See also Martin Murray's discovery and analysis of the military service records of Bill and Arthur Giggie, "Responding Kisses: New Evidence

about the Origins of 'Vigil Strange I Kept on the Field One Night,'" *Walt Whitman Quarterly Review* 25, no. 4 (Spring 2008): 192–97.

68. Roper, *Now the Drum of War*, 279.

69. Joseph Cady describes "Vigil Strange" as an attempt to put into words, in the absence of a cultural lexicon for same-sex desire, the intimacy between soldiers. Joseph Cady, "*Drum-Taps* and Male Homosexual Literature," in *Walt Whitman: Here and Now*, ed. Joann P. Krieg (Westport, CT: Greenwood, 1985), 49–59.

70. Whitman, *LG 1867*, 59a.

71. For Whitman's full correspondence with Sawyer, see Whitman, *Corr.*, 1:90–94. For further analysis of their relationship, see Epstein, *Lincoln and Whitman*, 136–39; Katz, *Love Stories*, 152–53; Loving, *Walt Whitman: Song of Himself*, 269–78; and Roper, *Now the Drum of War*, 205–7.

72. Walt Whitman, draft letter to Thomas P. Sawyer, April 26, 1863, Berg Collection, New York Public Library. This letter is written on the back of a flyer for the United States Christian Commission requesting donations similar to the items Whitman procured for Sawyer: "Clothing is greatly needed, such as cotton shirts, woolen shirts, woolen socks, flannel drawers, woolen drawers, surgical shirts of the same pattern, but tied with tapes down the sides and sleeves, instead of being sewed."

73. Walt Whitman, draft letter to Thomas P. Sawyer, May 27, 1863, Berg Collection.

74. Whitman, draft letter to Thomas P. Sawyer, April 26, 1863, Berg Collection.

75. Sawyer may have had a proxy draft the letter or assist with its composition (Whitman, *Corr.*, 1:90).

76. Thomas Sawyer, autographed letter signed to Walt Whitman, January 21, 1864, Berg Collection.

77. Walt Whitman, draft letter to Thomas P. Sawyer, July 1863, Berg Collection.

78. Whitman, draft letter to Thomas P. Sawyer, April 26, 1863, Berg Collection.

79. Whitman, *LG 1867*, 43a.

80. Whitman, "Dirge for Two Veterans," in *LG 1867*, 22a.

81. Whitman, *Corr.*, 1:127–29.

82. Price, "Debris," 59–80.

83. Bradford, *Communities of Death*, 129.

84. Whitman, *MDW*, 48.

85. Walt Whitman, "A Carol of Harvest, for 1867," *Galaxy* 4 (September 1867): 605–9. Whitman revised the poem for "Passage to India" (1871). After further revisions, it was retitled "The Return of the Heroes" in *Leaves of Grass* (1881–82). See "Poems in Periodicals," *The Walt Whitman Archive*. I have chosen to analyze this initial version because of its relevance to the theme of war as violent harvest, and its historical proximity to the war's conclusion.

86. Whitman, "A Carol of Harvest," 605.

87. Whitman, "A Carol of Harvest," 605.

88. Whitman, "A Carol of Harvest," 607.

89. Robert Leigh Davis, *Whitman and the Romance of Medicine*, 19.

90. Whitman, *LG 1891*, 407.

91. Whitman, *MDW*, 57.

92. Walt Whitman, "Pensive and Faltering," in *LG 1881*, 346.

93. Roper, *Now the Drum of War*, 336.

94. Loving, *Walt Whitman: Song of Himself*, 285–86.

95. "Biography," *Walt Whitman Archive.*

96. Moon, *Disseminating Whitman,* 200.

97. Of the seventy-one *Drum-Taps* and *Sequel* poems, only thirty-eight were retained in the clusters "Drum-Taps" and "Memories of President Lincoln" in the 1881 edition (Loving, *Walt Whitman: Song of Himself*, 286). Loving observes the symmetry between the suture of the *Sequel* to *Drum-Taps* and the integration of *Drum-Taps* into the 1867 *Leaves:* "Walt thought the 1867 *Leaves of Grass* was possibly his last edition (he thought the same thing about his new edition in 1871). But no sooner was it out than he decided to add more poems, though not new ones, to a second issue in the winter of 1867. Just as he had sewn the "Sequel" into *Drum-Taps,* he now sewed the 1865 texts of *Drum-Taps* with its "Sequel" into the new *Leaves* . . . . The fourth issue is therefore known as "The Workshop Edition" because of its heavy revisions" (*Walt Whitman: Song of Himself*, 315).

98. Whitman, *MDW*, 88.

99. Most recently, Martha Hodes has authored a definitive text on Lincoln's death pageant: *Mourning Lincoln* (New Haven, CT: Yale University Press, 2015).

100. Roper, *Now the Drum of War*, 343.

101. John R. Neff, *Honoring the Civil War Dead: Commemoration and the Problem of Reconciliation* (Lawrence: University of Kansas Press, 2005), 75.

102. Edward J. Russo and Curtis R. Mann, *Oak Ridge Cemetery* (Chicago: Arcadia, 2009), 21.

103. Blight, "Decoration Days," 94.

104. Neff, *Honoring the Civil War Dead*, 74.

105. For a detailed account of Whitman's interactions with Lincoln in Washington, see Epstein, *Lincoln and Whitman*, 160–64, 172–73, 177–80.

106. Whitman, *MDW*, 39–40.

107. Whitman, *MDW*, 40–41. An early draft of this passage was written on the back of a draft letter to Thomas Sawyer, dated July 1863, Berg Collection, New York Public Library.

108. Whitman, *MDW*, 48. On Whitman's fascination with Lincoln's face, see Alan Trachtenberg, "Lincoln's Smile: Ambiguities of the Face in Photography," in *Lincoln's Smile and Other Enigmas* (New York: Hill and Wang, 2007), 69–86.

109. Whitman, *MDW*, 39.

110. Whitman, *MDW*, 40.

111. Lincoln had two life masks made, first in April 1860 and then in February 1865. Observing the severity of the changes to Lincoln's face between the two masks, Lincoln's secretary, John Hay, wrote: "Under this frightful ordeal his demeanor and disposition changed. . . . The first is . . . a face full of life, of energy, of vivid aspiration. The other is so sad and peaceful in its infinite repose that the famous sculptor Augustus Saint-Gaudens insisted, when he first saw it, that it was a death-mask." Michael Burlingame, ed., *At Lincoln's Side: John Hay's Civil War Correspondence and Selected Writings* (Carbondale: Southern Illinois University Press, 2000), 139–40.

112. My reading of "Lilacs" has been influenced by Max Cavitch, chapter 6, "Retrievements Out of the Night," in *American Elegy,* 233–85; Michael Moon, "Uncanny Survivals: The Erotics of 'When Lilacs Last in the Dooryard Bloom'd,'" in *Disseminating*

Whitman, 215–21; Peter M. Sacks's epilogue to *English Elegy: Studies in the Genre from Spenser to Yeats* (Baltimore: Johns Hopkins University Press, 1985), 313–18; and Michael Warner, "Civil War Religion and Whitman's *Drum-Taps*," in *Walt Whitman: Where the Future Becomes Present*, eds. David Haven Blake and Michael Robertson (Iowa City: University of Iowa Press, 2008), 81–91.

113. Folsom and Price, *Re-Scripting Walt Whitman*, 91–92.

114. Whitman, *MDW*, 83.

115. Roper, *Now the Drum of War*, 339.

116. Max Cavitch positions "When Lilacs Last in the Door-Yard Bloom'd" as an elegiac rendering of the "fecundity of liminal spaces" within "a secret culture of desire" (*American Elegy*, 245, 256).

117. Whitman, *LG 1867*, 4b.

118. Whitman, *LG 1867*, 4b.

119. Whitman, *LG 1867*, 5b.

120. Whitman, *MDW*, 40.

121. Gregory Eiselein has linked "When Lilacs Last in the Door-Yard Bloom'd" to the genre of consolation poems. *Literature and Humanitarian Reform in the Civil War Era* (Bloomington: Indiana University Press, 1996).

122. Benjamin Franklin Morris, *Memorial Record of the Nation's Tribute to Abraham Lincoln* (Washington: W. H. & O. H. Morrison, 1896), 41–42.

123. Laderman, *Sacred Remains*, 159.

124. Laderman, *Sacred Remains*, 159.

125. Prior to the Civil War, embalming was not widely practiced; its chief use was the preservation of cadavers for dissection, and it was therefore scarred by that association (Faust, *Republic of Suffering*, 92; Keller Day, "Death and Dying," 277; Laderman, *Sacred Remains*, 3, 167–69). While the publication of graphic details of the president's embalming and autopsy was unprecedented, the autopsy of a public figure differed substantially from experimental dissection. Dismemberment of the corpse conflicted with nineteenth-century religious beliefs about resurrection, theoretically casting the subject into a form of purgatory. On the other hand, an autopsy marked the deceased as important enough to warrant investigation into the cause of death. See Nystrom, "Bioarchaeology of Structural Violence and Dissection," 1–15; Carlina De la Cova, "Patterns of Trauma and Violence in 19th-Century-Born African American and Euro-American Females," *International Journal of Paleopathology* 2, no. 2–3 (2012): 61–68.

126. Whitman, *MDW*, 157.

127. Benjamin Franklin Morris, *Memorial Record of the Nation's Tribute to Abraham Lincoln* (Washington, DC: W. H. & O. H. Morrison, 1896), 41.

128. Laderman, *Sacred Remains*, 160.

129. Laderman, *Sacred Remains*, 160.

130. Laderman, *Sacred Remains*, 160–61.

131. Laderman, *Sacred Remains*, 163.

132. Whitman, *LG 1867*, 5b.

133. Whitman, "A Carol of Harvest," 605.

134. Whitman, *LG 1867*, 7b.

135. Blight, "Decoration Days," 94, 100.

136. Neff, *Honoring the Civil War Dead*, 80. For more on Whitman's engagement with memorial artifacts of Lincoln, see Cavitch's analysis of Whitman's attachment to a *carte-de-visite* photograph of a lithograph published shortly after the assassination, "a strange little Washington-Lincoln photo. It represents Lincoln as being welcomed into the cloudlands and throwing his arms about Washington" (Cavitch, *American Elegy*, 233; Traubel, *With Walt Whitman in Camden*, 3:134–35).

137. Neff, *Honoring the Civil War Dead*, 80.

138. Keith Davis, "A Terrible Distinctness," 166.

139. In *Secure the Shadow: Death and Photography in America*, Jay Ruby incorrectly refers to "L. L. Rogers" as "claim[ing]" to specialize in skeleton leaf memorial stereographs (1995, p. 141). In his essay, '"A Terrible Distinctness': Photography of the Civil War Era," Keith F. Davis also misrepresents Rogers, attributing a reproduced image of her work, titled *Lincoln, Skeleton Leaves,* to "J. L. Rogers" (166) even though a copyright declaration "by I. L. Rogers" appears directly under the name of the stereographer, John P. Soule, on the original image.

140. Irene L. Rogers, Patent No. 211, 054, December 17, 1878, United States Patent Office, http://www.uspto.gov/index.jsp.

141. Laderman, *Sacred Remains*, 159.

142. This title references another Lincoln skeleton leaves stereograph (c. 1865–68). The artist who created the leaves arrangement is not identified. Copies are held at the Getty Museum and Art Institute of Chicago. "Lincoln Memorial. Dissected Leaves," Edward and Henry T. Anthony & Co., New York, undated.

143. Whitman, *LG 1867*, 11b.

144. Whitman, *LG 1855*, 16.

145. Whitman, *LG 1881*, 346.

146. Whitman, "A Carol of Harvest," 605.

Chapter Five

1. Loving, *Walt Whitman: Song of Himself*, 479. Alternatively, Roper suggests that he may have been infected by his sister-in-law, Mattie, whose death was likely the result of tuberculosis (*Now the Drum of War*, 77–79).

2. "Walt Whitman," *New York Mirror*, April 1, 1892.

3. Whitman, *MDW*, 30.

4. Traubel, *With Walt Whitman in Camden*, 3:581–82.

5. Quoted in Loving, *Walt Whitman: Song of Himself*, 461.

6. Traubel, *With Walt Whitman in Camden*, 3:581–82.

7. Loving, *Walt Whitman: Song of Himself*, 460.

8. Loving, *Walt Whitman: Song of Himself*, 458.

9. Whitman, *LG 1855*, 17.

10. Quoted in Kaplan, *Walt Whitman: A Life*, 52.

11. Quoted in Kaplan, *Walt Whitman: A Life*, 52.

12. William Sloane Kennedy quoted in Michael Robertson, *Worshipping Walt: The Whitman Disciples* (Princeton, NJ: Princeton University Press, 2008), 6–7. Traubel's fiancé, Anne Montgomerie, was also touched by Whitman's magnetism. She wrote

of the day of their meeting, "I felt a strange power radiating from him with so much force that I felt pushed away." She regularly visited the poet, and the couple were married in Whitman's bedroom (Loving, *Walt Whitman: Song of Himself*, 478).

13. Robertson, *Worshipping Walt*, 7.

14. See Matt Cohen, ed., introduction to *With Walt Whitman in Camden*, available on *The Walt Whitman Archive*, http://www.whitmanarchive.org/criticism/disciples/traubel/introduction.html.

15. Whitman, *Specimen Days*, 8.

16. Cohen, introduction.

17. Quoted in Cohen, introduction.

18. Traubel, *With Walt Whitman in Camden*, 9:604–5.

19. Traubel, *With Walt Whitman in Camden*, 9:261.

20. The "Good Grey Poet" was not yet an uncontroversial figure at home. As the *New York Mirror* wrote in Whitman's obituary, Whitman's "fame grew wider and wider abroad. In England there would have been astonishment had it been known how poor and neglected he was at home." Again, there is an element of sensationalism to this account. While Whitman's legacy remained haunted by puritanical American responses to the more erotic elements of *Leaves of Grass*, he was hardly "neglected." See Robertson's introduction in *Worshipping Walt: The Whitman Disciples* (1–14).

21. "The Aged Bard Passed Away Peacefully Last Night," *Philadelphia Record,* March 27, 1892.

22. "The Dead Poet's Brain: A Report that He Bequeathed It to the Anthropometric Society," *The New York World,* March 28, 1892. Hereafter "Dead Poet's Brain."

23. This obituary was sourced from a clipping in the Harned Collection of the Papers of Walt Whitman at the Library of Congress. The print source of the nineteenth-century article is *The Philadelphia Bureau of Press Clippings.* However, the original newspaper citation has not survived.

24. See Kaplan, *Walt Whitman: A Life*, 52.

25. Traubel, *With Walt Whitman in Camden*, 9:601.

26. Traubel, *With Walt Whitman in Camden*, 9:602.

27. John Caspar Lavater, *Essays on Physiognomy: Designed to Promote the Knowledge and Love of Mankind* (London: William Tess, 1858), 149.

28. Loving, *Walt Whitman: Song of Himself*, 460

29. See Amy Beth Werbel, chapter 3, "Body Casts and 'Anatomical Eyes,'" in *Thomas Eakins: Art, Medicine and Sexuality in Nineteenth-Century Philadelphia* (New Haven, CT: Yale University Press, 2007), 53–86.

30. Werbel, "Body Casts," 72.

31. Traubel, *With Walt Whitman in Camden*, 9:602–3, 606.

32. Traubel, *With Walt Whitman in Camden*, 9:604–5.

33. Traubel, *With Walt Whitman in Camden*, 9:601.

34. Whitman, *Corr.*, 1:127–29.

35. "Dead Poet's Brain."

36. Traubel, *With Walt Whitman in Camden*, 9:603–4.

37. "Dead Poet's Brain."

38. Nystrom, "Bioarchaeology of Structural Violence and Dissection," 11.

39. "Dead Poet's Brain."

40. Brian Burrell, "The Strange Fate of Whitman's Brain," *Walt Whitman Quarterly Review* 20 (2003): 107–33; Sheldon Lee Gosline, "'I Am a Fool': Dr. Henry Cattell's Private Confession about What Happened to Whitman's Brain," *Walt Whitman Quarterly Review* 31 (2014): 158–62.

41. Burrell, "Strange Fate of Whitman's Brain," 109; Gosline, "'I Am a Fool,'" 159.

42. Burrell, "Strange Fate of Whitman's Brain," 123.

43. Burrell, "Strange Fate of Whitman's Brain," 107.

44. Burrell, "Strange Fate of Whitman's Brain," 110.

45. Burrell, "Strange Fate of Whitman's Brain," 123.

46. Burrell, "Strange Fate of Whitman's Brain," 111. While Traubel and others made public statements of outrage, no formal action was taken.

47. Quoted in Burrell, "Strange Fate of Whitman's Brain," 111.

48. Traubel, *With Walt Whitman in Camden*, 9:605; Burrell, "Strange Fate of Whitman's Brain," 121.

49. Gosline, "'I Am a Fool,'" 157–58.

50. Gosline, "'I Am a Fool,'" 159.

51. Gosline, "'I Am a Fool,'" 124.

52. Burrell, "Strange Fate of Whitman's Brain," 110.

53. Burrell, "Strange Fate of Whitman's Brain," 110.

54. Hunt, *Address Upon the Late Joseph Leidy*, 27–29.

55. Traubel, *With Walt Whitman in Camden*, 9:604.

56. Loving, *Walt Whitman: Song of Himself*, 479.

57. Traubel, *With Walt Whitman in Camden*, 9:604–5.

58. Traubel, *With Walt Whitman in Camden*, 9:604–5.

59. Whitman, *PP*, 1214.

60. Whitman, *LG 1855*, 26.

61. Abraham and Torok, *Shell and the Kernel*, 118.

62. Whitman, *SD*, 8.

63. Traubel, *With Walt Whitman in Camden*, 9:613.

64. Traubel, *With Walt Whitman in Camden*, 9:261.

65. Ed Folsom, "Horace Traubel," in *Walt Whitman: An Encyclopedia*, eds. J. R. LeMaster and Donald D. Kummings, available on *The Walt Whitman Archive.*

66. Gary Schmidgall, ed., *Conserving Walt Whitman's Fame: Selections from Horace Traubel's "Conservator," 1891–1919* (Iowa City: University of Iowa Press, 2006), xviii.

67. Folsom, "Horace Traubel."

68. Ed Folsom, *Whitman Making Books/Books Making Whitman: A Catalog and Commentary,* available on *The Walt Whitman Archive.*

69. Quoted in Folsom, *Whitman Making Books.*

70. Folsom, *Whitman Making Books.* See also Michael Moon, ed., *The Norton Critical Edition of Leaves of Grass* (New York: Norton, 2002), 425.

71. Folsom, *Whitman Making Books.*

72. Folsom, *Whitman Making Books.*

73. Folsom, *Whitman Making Books.*

74. Loving, *Walt Whitman: Song of Himself*, 476.

75. Folsom, *Whitman Making Books.*

76. Loving, *Walt Whitman: Song of Himself*, 476.

77. Quoted in Reynolds, *Walt Whitman's America*, 586.

78. Quoted in Reynolds, *Walt Whitman's America*, 586.

79. Reynolds, *Walt Whitman's America*, 587.

80. Reynolds, *Walt Whitman's America*, 589.

81. Whitman, *LG 1891*, 407.

82. J. M. Coetzee, "Walt Whitman," in *Inner Workings: Literary Essays 2000–2005* (New York: Random House, 2007), 186.

83. Whitman, *LG 1891*, 407.

84. Whitman, *LG 1891*, 407.

85. Traubel, *With Walt Whitman in Camden*, 8:58.

86. Whitman, *LG 1891*, 408.

87. Whitman, *LG 1891*, 407.

88. Whitman, *LG 1891*, 407.

89. Whitman, *LG 1891*, 408.

90. Whitman, *LG 1891*, 410.

91. Whitman, *LG 1891*, 410.

92. Whitman, *LG 1891*, 410.

93. "Whitman's Farewell," *The New York Tribune*, August 16, 1891. See "An Album of Portraits" in *The Norton Critical Edition of Leaves of Grass*, ed. Michael Moon, for a reproduction of the photographic frontispiece to *Good-Bye My Fancy* (1891).

94. Whitman, *LG 1891*, 408.

95. *Oxford English Dictionary*.

96. *Oxford English Dictionary*. See Robert Creeley's essay, "Reflections on Whitman in Age," for an interesting etymological reading of the word "fancy" (65), in *On Earth* (Berkeley: University of California Press, 2006), 59–87.

97. Whitman, *LG 1891*, 408, emphasis mine.

98. Whitman, *LG 1891*, 1. Pages 423–38 refer to Whitman's nostalgic prose essay "A Backward Glance O'er Travel'd Roads," which immediately followed "Good-Bye My Fancy" in the final edition of *Leaves of Grass*.

99. Derrida, *Memoires pour Paul de Man*, 122.

100. Derrida, *Work of Mourning*, 170.

101. Whitman, *LG 1867*, 7b; Kaplan, *Walt Whitman: A Life*, 5.

102. Reynolds, *Walt Whitman's America*, 571.

103. Reynolds, *Walt Whitman's America*, 572.

104. Loving, *Walt Whitman: Song of Himself*, 473.

105. See Loving, *Walt Whitman: Song of Himself*, 473; Reynolds, *Walt Whitman's America*, 572; Traubel, *With Walt Whitman in Camden*, 9:210–12. On correlations between Whitman's final text and his tomb, see Amanda Gailey, "Whitman's Shrines," in *Proofs of Genius* (Ann Arbor: University of Michigan Press, 2015).

106. See Whitman, *Corr.*, 5:95. See also Sarah Ferguson-Wagstaffe, "'Points of Contact': Blake and Whitman," in *Sullen Fires across the Atlantic: Essays in Transatlantic Romanticism*, ed. Lance Newman, Chris Koenig-Woodyard, and Joel Pace, *Romantic Circles: Praxis Series*, February 2010, http://www.rc.umd.edu/praxis/sullenfires/sfw/sfw_essay.html.

107. Whitman, *LG 1867*, 7b.

108. Ferguson-Wagstaffe, "Points of Contact."

109. Ferguson-Wagstaffe, "Points of Contact."

110. "Whitman's Tomb: Description of the Mausoleum High in Harleigh's Hillside," *Camden Courier*, March 29, 1892. Hereafter "Whitman's Tomb."

111. Kaplan, *Walt Whitman: A Life*, 50.

112. "Whitman's Tomb"; Kaplan, *Walt Whitman: A Life*, 50.

113. "Whitman's Tomb"; Kaplan, *Walt Whitman: A Life*, 50; Reynolds, *Walt Whitman's America*, 572.

114. As David Reynolds has established, Whitman became entangled in financial turmoil regarding funding for the tomb (*Walt Whitman's America*, 572). For further details see Traubel, *With Walt Whitman in Camden*, 9:25–26, 38, 85–87; and Kaplan, *Walt Whitman: A Life*, 48.

115. Kaplan, *Walt Whitman: A Life*, 50–51.

116. Kaplan, *Walt Whitman: A Life*, 51.

117. Kaplan, *Walt Whitman: A Life*, 50–51.

118. Whitman quoted in Reynolds, *Walt Whitman's America*, 586.

119. Whitman, *LG 1855, 56.*

120. Whitman, *LG 1855*, 56.

121. Whitman, *LG 1855*, 56.

122. Whitman, *MDW*, 16.

123. Quoted in Kaplan, *Walt Whitman: A Life*, 51.

124. Kaplan, *Walt Whitman: A Life*, 51.

125. Whitman, *MDW*, 56.

126. Kaplan, *Walt Whitman: A Life*, 51.

127. Whitman, *MDW*, 57.

128. Whitman, "A Carol of Harvest," 605.

129. "Whitman's Tomb."

130. "Whitman's Tomb."

131. Traubel, *With Walt Whitman in Camden*, 9:123.

132. Whitman conceived of *Leaves of Grass* as a "Great Construction of the New Bible." W. C. Harris, "Whitman's *Leaves of Grass* and the Writing of a New American Bible," *Walt Whitman Quarterly Review* 16 (1999): 172–90.

133. Traubel, *With Walt Whitman in Camden*, 9:601.

134. Traubel, *With Walt Whitman in Camden*, 9:617.

135. Reynolds, *Walt Whitman's America*, 589.

136. Reynolds, *Walt Whitman's America*, 589.

137. Loving, *Walt Whitman: Song of Himself*, 480.

WORKS CITED

Abbas, Ackbar. "Walter Benjamin's Collector: The Fate of Modern Experience." *New Literary History* 20, no. 1 (1988): 217–37.

Abraham, Nicholas, and Maria Torok. *The Shell and the Kernel: Renewals of Psychoanalysis*. Ed. and trans. Nicholas T. Rand. Chicago: University of Chicago Press, 1994.

———. *The Wolf Man's Magic Word: A Cryptonymy*. Trans. Nicholas Rand. Minneapolis: University of Minnesota Press, 1986.

"The Aged Bard Passed Away Peacefully Last Night." *Philadelphia Record*. March 27, 1892.

Albert A. Berg Collection of the Papers of Walt Whitman. New York Public Library.

Allen, Francis D. *Documents and Facts, Showing the Fatal Effects of Interments in Populous Cities*. New York: F. D. Allen, 1822.

Ames, Mary Clemmer. *Ten Years in Washington: Life and Scenes in the Nation's Capital, as a Woman Sees Them*. Hartford CT: Worthington, 1873.

Anderson, Donald Lee, and Anderson Godfrey Tryggve. "Nostalgia and Malingering in the Military during the Civil War." *Perspectives in Biology and Medicine* 28 (1984): 156–66.

Aries, Philippe. *Western Attitudes towards Death: From the Middle Ages to the Present*. Baltimore: Johns Hopkins University Press, 1975.

Arrojo, Rosemary. "Translation, Transference, and the Attraction to Otherness: Borges, Menard, Whitman." *Diacritics* 34, no. 3 (2004): 1047–82.

Aspiz, Harold. "Science and Pseudoscience." In *A Companion to Walt Whitman*, ed. Donald D. Kummings, 216–33. Oxford: Blackwell, 2009.

———. *So Long!: Walt Whitman's Poetry of Death*. Tuscaloosa: University of Alabama Press, 2004.

———. "Specimen Days: The Therapeutics of Sun-Bathing." *Walt Whitman Quarterly Review* 1 (1983): 48–50.

Barbian, Lenore, Paul S. Sledzik, and Jeffrey Reznick. "Remains of War: Walt Whitman, Civil War Soldiers, and the Legacy of Medical Collections." *Museum History Journal* 5, no. 1 (2012): 7–28.

Barthes, Roland. *Camera Lucida: Reflections on Photography*. Trans. Richard Howard. New York: Hill and Wang, 1982.

Bartholow, Robert. *Manual of Instructions for Enlisting and Discharging Soldiers*. Philadelphia: Lippincott, 1863.

Bauman, Richard A. *Human Rights in Ancient Rome.* London: Routledge, 2000.

Bayley, John. "Songs of a Furtive Self: Whitman." In *Selected Essays*, 1–7. Cambridge: Cambridge University Press, 1984.

Beach, Christopher. *The Politics of Distinction: Whitman and the Discourses of Nineteenth-Century America.* Athens: University of Georgia Press, 1996.

Beaver, Joseph. *Walt Whitman—Poet of Science*. New York: Octagon Books, 1974.

Benjamin, Walter. *The Arcades Project*. Ed. Rolf Tiedemann, trans. Howard Eiland and Kevin McLaughlin. Cambridge, MA: Harvard University Press, 1999.

———. *One Way Street and Other Writings*. Trans. Edmund Jephcott and Kingsley Shorter. 1979. London: Verso, 1992.

Benkard, Ernst. *Undying Faces: A Collection of Death Masks.* Trans. Margaret M. Green. London: Hogarth, 1929.

Bertolini, Vincent J. "'Hinting' and 'Reminding': The Rhetoric of Performative Embodiment in *Leaves of Grass*." *English Literary History* 69, no. 4 (2002): 1047–82.

Blake, David Haven. *Walt Whitman and the Culture of American Celebrity*. New Haven, CT: Yale University Press, 2006.

Blasing, Mutlu Konuk. "Whitman's 'Lilacs' and the Grammars of Time." *PMLA* 97, no. 1 (1982): 31–39.

Blight, David W. "Decoration Days: The Origins of Memorial Day in the North and South." In *The Memory of the Civil War in American Culture*, eds. Alice Fahs and Joan Waugh, 94–129. Chapel Hill: University of North Carolina Press.

Bollet, Alfred. *Civil War Medicine: Challenges and Triumphs.* Tucson, AZ: Galen, 2002.

Boss, Pauline. *Ambiguous Loss: Learning to Live with Unresolved Grief.* Cambridge, MA: Harvard University Press, 1999.

Boym, Svetlana. *The Future of Nostalgia.* New York: Basic Books, 2001.

Bradford, Adam. *Communities of Death: Whitman, Poe and the American Culture of Mourning*. Columbia: University of Missouri Press, 2014.

———. "Re-collecting Soldiers: Walt Whitman and the Appreciation of Human Value." *Walt Whitman Quarterly Review* 27 (2010): 127–152.

Bradley, Sculley, Harold W. Blodgett, Arthur Golden, and William White, eds. "Cluster Arrangements in *Leaves of Grass*." In *Leaves of Grass: A Textual Variorum of the Printed Poems*, vol. 1: lvii–lxxvii. New York: New York University Press, 1980.

Brault, Pascale-Anne. "To Reckon with the Dead: Jacques Derrida's Politics of Mourning." Editor's introduction. In *The Work of Mourning*, by Jacques Derrida, eds. Pascale-Anne Brault and Michael Naas. Chicago: University of Chicago Press, 2001.

Breitwieser, Mitchell Robert. *National Melancholy: Mourning and Opportunity in Classic American Literature.* Stanford, CA: Stanford University Press, 2007.

Brinton, John H. *Personal Memoirs of John H. Brinton, Major and Surgeon, U.S.V., 1861–1865*. Carbondale, IL: Shawnee Classics, 1996.

Browne, Janet. "Botany for Gentlemen: Erasmus Darwin and the 'Loves of the Plants.'" *Isis* 80, no. 4 (1989): 593–621.

Brugger, Peter. "The Phantom Limb in Dreams." *Conscious Cognition* 17, no. 4 (2008): 1279–80.

Buinicki, Martin T. *Walt Whitman's Reconstruction: Poetry and Publishing between Memory and History.* Iowa City: University of Iowa Press, 2011.

Burlingame, Michael, ed. *At Lincoln's Side: John Hay's Civil War Correspondence and Selected Writings.* Carbondale: Southern Illinois University Press, 2000.

Burrell, Brian. "The Strange Fate of Whitman's Brain." *Walt Whitman Quarterly Review* 20 (Winter/Spring 2003): 107–33.

Burroughs, John. *Notes on Walt Whitman as Poet and Person*. New York: American News, 1867.

Butler, Judith. *Antigone's Claim: Kinship between Life and Death.* New York: Columbia University Press, 2000.

———. *Precarious Life: The Powers of Mourning and Violence*. New York: Verso, 2004.

———. *The Psychic Life of Power: Theories in Subjection*. Stanford, CA: Stanford University Press, 1997.

Cady, Joseph. "*Drum-Taps* and Male Homosexual Literature." In *Walt Whitman: Here and Now*, ed. Joann P. Krieg, 49–59. Westport, CT: Greenwood, 1985.

Cannavo, Peter F. *The Working Landscape: Founding, Preservation, and the Politics of Place*. Cambridge, MA: MIT Press, 2007.

Caruth, Cathy. *Unclaimed Experience: Trauma, Narrative and History.* Baltimore: Johns Hopkins University Press, 1996.

Casey, Edward S. "The World of Nostalgia." *Man and World* 20 (1987): 361–84.

Catalogue of the Surgical Section of the US Army Medical Museum. Washington, DC: Government Printing Office, 1866.

Cavitch, Max. *American Elegy: The Poetry of Mourning from the Puritans to Whitman.* Minneapolis: University of Minnesota Press, 2007.

Cervetti, Nancy. *S. Weir Mitchell: Philadelphia's Literary Physician, 1829–1914*. University Park: Pennsylvania State University Press, 2012.

Charles E. Feinberg Collection of the Papers of Walt Whitman. Library of Congress.

Chisholm, Julian John. *A Manual of Military Surgery for the Use of the Surgeons in the Confederate Army*, 3rd ed. Columbia, SC: Evans and Cogswell, 1864.

Clewell, Tammy. "Mourning beyond Melancholia: Freud's Psychoanalysis of Loss." *Journal of the American Psychoanalytic Association* 52, no. 1 (2004): 43–67.

Coetzee, J. M. "Walt Whitman." In *Inner Workings: Literary Essays 2000–2005*, 174–88. New York: Random House, 2007.

Cohen, Matt. "Transgenic Deformation: Literary Translation and the Digital Archive." In *The Walt Whitman Archive*, eds. Ed Folsom and Kenneth M. Price.

Cole, Hubert. *Things for the Surgeon: A History of the Body-Snatchers.* London: Heineman, 1964.

Connor, J. T. H. and Michael Rhode. "Shooting Soldiers: Civil War Medical Images, Memory, and Identity in America." *Invisible Culture* 5 (2003). Web.

Coviello, Peter. *Intimacy in America: Dreams of Affiliation in Antebellum Literature.* Minneapolis: University of Minnesota Press, 2005.

———. "Intimate Nationality: Anonymity and Attachment in Whitman." *American Literature* 73, no. 1 (2001): 85–119.

———. Preface. "Whitman at War." In *Memoranda during the War*, by Walt Whitman, ix–liv. 1875–76. Oxford: Oxford University Press, 2004.

Cowley, Malcom, ed. "Introduction." 1959. In *Leaves of Grass*, by Walt Whitman, vii–xxxvii. 1855. New York: Penguin, 1986.

Crabtree, Adam. *From Mesmer to Freud: Magnetic Sleep and the Roots of Psychological Healing.* New Haven, CT: Yale University Press, 1993.

Crawford, Cassandra S. *Phantom Limb: Amputation, Embodiment and Prosthetic Technology.* New York: New York University Press, 2014.

Creeley, Robert. "Reflections on Whitman in Age." In *On Earth*, 59–87. Berkeley: University of California Press, 2006.

Crocq, Marc-Antoine, and Louis Crocq. "From Shell Shock and War Neurosis to Posttraumatic Stress Disorder: A History of Psychotraumatology." *Dialogues in Clinical Neuroscience* 2, no. 1 (2000): 47–55.

Darrah, William C. *The World of Stereographs*. Gettysburg, PA: Darrah, 1977.

Darwin, Charles. *The Origin of Species*. Ed. Gillian Beer. Oxford: Oxford University Press, 1996.

Davenport, Guy. "Whitman." In *The Geography of the Imagination: Forty Essays,* 68–79. Jaffrey, NH: Godine, 1997.

Davis, Fred. *Yearning for Yesterday: A Sociology of Nostalgia*. New York: Free, 1979.

Davis, Keith. "'A Terrible Distinctness': Photography in the Civil War Era." In *Photography in Nineteenth-Century America*, ed. Martha A. Sandweiss, 130–79. New York: Harry N. Abrams, 1991.

Davis, Robert Leigh. *Whitman and the Romance of Medicine*. Berkeley: University of California Press, 1997.

Dawes, James. *The Language of War: Literature and Culture in the U.S. from the Civil War through World War II.* Cambridge, MA: Harvard University Press, 2002.

Day, Nicole Keller. "Death and Dying." In *Civil War America: A Social and Cultural History*, eds. Maggi M. Morehouse and Zoe Trodd, 275–84. New York: Routledge, 2013.

"The Dead Poet's Brain: A Report That He Bequeathed It to the Anthropometric Society." *The New York World.* March 28, 1892.

Dean, Eric T. *Shook Over Hell: Post-Traumatic Stress, Vietnam, and the Civil War.* Cambridge, MA: Harvard University Press, 1997.

De la Cova, Carlina. "Patterns of Trauma and Violence in Nineteenth-Century-Born African American and Euro-American Females. *International Journal of Paleopathology* 2, no. 2–3 (2012): 61–68.

Derrida, Jacques. *Aporias*. Trans. Thomas Dutoit. Stanford, CA: Stanford University Press, 1993.

———. "*Fors*: The Anglish Words of Nicholas Abraham and Maria Torok." Trans. Barbara Johnson. In *The Wolf Man's Magic Word: A Cryptonymy*, by Nicholas Abraham and Maria Torok, trans. Nicholas Rand, ix–il. Minneapolis: University of Minnesota Press, 1986.

———. *Memoires pour Paul de Man*. Trans. Cecile Lindsay, Jonathan Culler, Eduardo Cadava, and Peggy Kamuf. New York: Columbia University Press, 1989.

———. *Of Grammatology*. Trans. Gayatri Chakravorty Spivak. Baltimore: Johns Hopkins University Press, 1974.

———. *Of Hospitality*. Trans. Rachel Bowlby. Stanford, CA: Stanford University Press, 2000.

———. *On Touching—Jean-Luc Nancy.* Trans. Christine Irizarry. Stanford, CA: Stanford University Press, 2005.

———. *Spectres of Marx: The State of Debt, The Work of Mourning, and The New International.* Trans. Peggy Kamuf. New York: Routledge, 1994.
———. *The Work of Mourning.* Eds. Pascale-Anne Brault and Michael Naas. Chicago: University of Chicago Press, 2001.
Dollimore, Jonathan. *Death, Desire and Loss in Western Culture.* New York: Routledge, 1998.
Douglas, Ann. *The Feminization of American Culture.* New York: Alfred A. Knopf, 1977.
Dufourmantelle, Anne. "Invitation." In *Of Hospitality*, by Jacques Derrida, trans. Rachel Bowlby. Stanford, CA: Stanford University Press, 2000.
Dufresne, Todd. *Tales from the Freudian Crypt: The Death Drive in Text and Context.* Stanford, CA: Stanford University Press, 2000.
Dunlap, David. "Security Concerns Raised about Memorial at Ground Zero." *The New York Times.* April 2, 2006.
Dwight, Thomas. "Anatomy Laws vs. Body Snatching." *Forum* 22 (1896): 493–502.
Earnest, Ernest. *S. Weir Mitchell: Novelist and Physician.* Philadelphia: University of Pennsylvania Press, 1950.
Eicher, David J. *The Longest Night: A Military History of the Civil War.* New York: Simon & Schuster, 2001.
Eiland, Howard, and Kevin McLaughlin. "Translator's Foreword." In *The Arcades Project*, by Walter Benjamin, trans. Howard Eiland and Kevin McLaughlin, ed. Rolf Tiedemann. Cambridge, MA: Harvard University Press, 1999.
Eiselein, Gregory. *Literature and Humanitarian Reform in the Civil War Era.* Bloomington: Indiana University Press, 1996.
Ellis, Edward Robb. *The Epic of New York City: A Narrative History.* New York: Carrol & Graf, 2004.
Ellsworth, Henry. "Letter from Patent Office Commissioner Henry Ellsworth to Senator John Ruggles, Dec. 18, 1840." National Portrait Gallery. Smithsonian Institution. http://www.npg.si.edu/exhibit/pob/.
Emerson, Ralph Waldo. *Emerson: Essays and Lectures.* New York: Penguin Literary Classics, 1983.
———. *The Journals and Miscellaneous Notebooks of Ralph Waldo Emerson.* 4 vols. Ed. Albert J. von Frank. Columbia: University of Missouri Press, 1989.
Epstein, Daniel Mark. *Lincoln and Whitman: Parallel Lives in Civil War Washington.* New York: Random House, 2004.
Erkkila, Betsy. *Whitman: The Political Poet.* New York: Oxford University Press, 1989.
Fabian, Ann. *The Skull Collectors: Race, Science and America's Unburied Dead.* Chicago: University of Chicago Press, 2010.
Farland, Maria. "Decomposing City: Walt Whitman's New York and the Science of Life and Death." *English Literary History* 74 (2007): 799–827.
Farr, Judith, with Louise Carter. *The Gardens of Emily Dickinson.* Cambridge, MA: Harvard University Press, 2004.
Faust, Drew Gilpin. *This Republic of Suffering: Death and the American Civil War.* New York: Alfred A. Knopf, 2008.
Feldman, Mark B. "Remembering a Convulsive War: Whitman's Memoranda during the War and the Therapeutics of Display." *Walt Whitman Quarterly Review* 23 (2005): 1–25.

Fenton, Elizabeth, and Valerie Rohy. "Whitman, Lincoln, and the Union of Men." *ESQ: A Journal of the American Renaissance* 55, no. 3 (2009): 237–67.
Ferguson-Wagstaffe, Sarah. "'Points of Contact': Blake and Whitman." In *Sullen Fires across the Atlantic: Essays in Romanticism*, eds. Lance Newman, Chris Koenig-Woodyard, and Joel Pace. November 2006. Romantic Circles Praxis Series, edited by Orrin Wang. http://www.rc.umd.edu/praxis/sullenfires/index.html.
Figg, Laurann, and Jane Farrell-Beck. "Amputation in the Civil War: Physical and Social Dimensions." *Journal of the History of Medicine and Allied Sciences* 48, no. 4 (1993): 454–75.
Finch, Caleb E. "Aging, Inflammation and the Body Electric." *Dædalus* 135 (2006): 68–76.
Flagg, Wilson. *Mount Auburn: Its Scenes, Its Beauties, and Its Lessons.* Boston: James Monroe, 1861.
Fletcher, Angus. *A New Theory for American Poetry: Democracy, the Environment, and the Future of Imagination.* Cambridge, MA: Harvard University Press, 2004.
Fluck, Hans, and Rita Jaspersen-Schib. *Medicinal Plants and Their Uses.* Trans. J. M. Rowson. London: W. Foulsham, 1976.
Folsom, Ed. "Lucifer and Ethiopia: Whitman, Race, and Poetics Before the Civil War and After." In *A Historical Guide to Walt Whitman*, ed. David S. Reynolds, 45–96. Oxford: Oxford University Press, 2000.
———. "Nineteenth-Century Visual Culture. In *A Companion to Walt Whitman*, ed. Donald D. Kummings, 272–90. Oxford: Blackwell, 2009.
———. "Poets of Compassion: Walt Whitman, Emily Dickinson and War." *Multitudes* (Summer 2003): 2–7.
———. "Walt Whitman and the Civil War: Making Poetry Out of Pain, Grief, and Mass Death." *Abaton* 2 (Fall 2008): 12–26.
———. *Whitman Making Books/Books Making Whitman: A Catalog and Commentary.* Iowa City, Iowa: Obermann Center for Advanced Studies, 2005. Sourced from *The Walt Whitman Archive.*
———. *Walt Whitman's Native Representations.* Cambridge: Cambridge University Press, 1994.
Folsom, Ed, and Kenneth M. Price, eds. *Re-Scripting Walt Whitman: An Introduction to His Life and Work.* New York: Wiley-Blackwell, 2005.
———. *The Walt Whitman Archive.* http://www.whitmanarchive.org/.
Fone, Byrne S. *Masculine Landscapes: Walt Whitman and the Homoerotic Text.* Carbondale: Southern Illinois University Press, 1992.
Foucault, Michel. *The Birth of the Clinic: An Archeology of Medical Perception.* New York: Routledge, 2003.
———. *The History of Sexuality.* Vol. 1. Trans. Robert Hurley. London: Penguin, 1998.
Franchetti, Michael A. "Trauma Surgery during the Civil War." *Southern Medical Journal* 86, no. 5 (1993): 553–56.
Frank, Adam. "Emily Dickinson and Photography." *Emily Dickinson Journal* 10, no. 2 (2001): 1–21.
———. "Phantom Limb: Silvan Tomkins and Affective Prosthetics." *Theory and Psychology* 17, no. 4 (2007): 515–28.
Frassanito, William A. *Antietam: The Photographic Legacy of America's Bloodiest Day.* New York: Charles Scribner's Sons, 1978.

Freemon, Frank R. *Gangrene and Glory: Medical Care during the Civil War.* London: Associated University Presses, 1998.

Freud, E. L., ed. *Letters of Sigmund Freud, 1873–1939.* Trans. Thomas Roberts. New York: The Other, 2003.

Freud, Sigmund. *Beyond the Pleasure Principle: On Metapsychology: The Theory of Psychoanalysis.* Eds. Albert Dickson and Angela Richards, trans. James Strachey, 275–338. New York: Penguin, 1991.

———. *The Interpretation of Dreams.* Trans. A. A. Brill. 1913. New York: Dover, 2015.

———. "Mourning and Melancholia." In *On Murder, Mourning, and Melancholia*, trans. Shaun Whiteside, 201–19. New York Penguin Modern Classics, 2005.

———. *Studies on Hysteria*. Trans. and ed. James Strachey. New York: Basic Books, 1957.

Gagarin, Michael, and Elaine Fantham, eds. *The Oxford Encyclopedia of Ancient Greece and Rome*. Vol. 1. Oxford: Oxford University Press, 2010.

Gailey, Amanda. *Proofs of Genius: Collected Editions from the American Revolution to the Digital Age.* Ann Arbor: University of Michigan Press, 2015.

Genoways, Ted. "Civil War Poems in 'Drum-Taps' and 'Memories of President Lincoln.'" In *A Companion to Walt Whitman*, ed. Donald D. Kummings, 522–39. Oxford: Blackwell, 2009.

———. *Walt Whitman and the Civil War: America's Poet during the Lost Years of 1860–1862*. Berkeley: University of California Press, 2009.

Gerhardt, Christine. *A Place for Humility: Whitman, Dickinson and the Natural World.* Iowa City: University of Iowa Press, 2014.

Gewirtz, Isaac. "I Am with [Some of] You." In *I Am with You: Walt Whitman's "Leaves of Grass" 1855–2005,* 10–47. New York: New York Public Library, 2005.

Gilman, Charlotte Perkins. *The Living of Charlotte Perkins Gilman: An Autobiography.* 1935. University of Wisconsin Press, 1963.

Gold, Moshe. "Serious Play: Derrida and Whitman in the Theory Classroom." *Symploke* 12, no. 1 (2004): 216–31.

Goler, Robert I. "Loss and the Persistence of Memory: 'The Case of George Dedlow' and Disabled Civil War Veterans." *Literature and Medicine* 23, no. 1 (2004): 160–83.

Goodall, Jane, and Christopher Lee, eds. "Introduction." In *Trauma and Public Memory,* 1–22. New York: Palgrave Macmillan Memory Studies Series, 2015.

Gosline, Sheldon Lee. "'I Am a Fool': Dr. Henry Cattell's Private Confession about What Happened to Whitman's Brain." *Walt Whitman Quarterly Review* 31 (2014): 158–62.

Griscom, John H. *The Sanitary Condition of the Laboring Population of New York, with Suggestions for Improvement*. New York: Harper and Bros., 1845. Hacker, J. David. "A Census-Based Count of the Civil War Dead." *Civil War History* 57, no. 4 (2011): 307–48.

Halligan, Peter W. "Phantom Limbs: The Body in Mind." *Cognitive Neuropsychiatry* 7, no. 3 (2002): 251–69.

Hallock, Thomas. "Male Pleasure and the Genders of Eighteenth-Century Botanic Exchange: A Garden Tour." *William and Mary Quarterly* 62, no. 4 (2005): 697–718.

Halperin, Edward C. "A Glimpse of Our Past: The Poor, the Black and the Marginalized as the Source of Cadavers in United States Anatomical Education." *Clinical Anatomy* 20, no. 5 (2007): 489–95.

Harris, W. C. "Whitman's *Leaves of Grass* and the Writing of a New American Bible." *Walt Whitman Quarterly Review* 16 (1999): 172–90.
Hasegawa, Guy R. *Mending Broken Soldiers: The Union and Confederate Programs to Supply Artificial Limbs*. Carbondale: Southern Illinois University Press, 2012.
Headley, J. T. *The Great Riots of New York 1712–1873*. New York: E. B. Treat, 1873.
Henderson, Desiree. "'What Is the Grass?': The Roots of Walt Whitman's Cemetery Meditation." *Walt Whitman Quarterly Review* 25, no. 3 (2008): 89–107.
Herman, J. "Phantom Limb: From Medical Knowledge to Folk Wisdom and Back." *Annals of Internal Medicine* 128 (1998): 76–78.
Herrington, Elrid. "Nation and Identity." In *A Companion to Walt Whitman*, ed. Donald D. Kummings, 122–35. Oxford: Blackwell, 2009.
Hodes, Martha. *Mourning Lincoln*. New Haven, CT: Yale University Press, 2015.
Hoffman, Julius. "Phantom Limb Syndrome: A Critical Review of Literature." *Journal of Nervous and Mental Disease* 119 (1954): 261–70.
Holmes, Oliver Wendell. "Doings of the Sunbeam." In *Soundings, from the Atlantic*, 266–67. Boston: Ticknor and Fields, 1866.
———. "The Human Wheel, Its Spokes and Felloes." *Atlantic Monthly*. May 1863, 567–80.
Hsu, David. "Walt Whitman: An American Civil War Nurse Who Witnessed the Advent of Modern American Medicine." *Archives of Environmental & Occupational Health* 65, no. 4 (2010): 238–39.
Humphrey, David C. "Dissection and Discrimination: The Social Origins of Cadavers in America, 1760–1915." *Bulletin of the New York Academy of Medicine* 49, no. 9 (1973): 819–27.
Hunt, William. *An Address Upon the Late Joseph Leidy*. Philadelphia: Collins, 1892.
Hutchinson, George. *The Ecstatic Whitman: Literary Shamanism and the Crisis of the Union*. Columbus: Ohio State University Press, 1986.
Hutchinson, George and David Drews. "*Specimen Days*." In *Walt Whitman: An Encyclopedia*, ed. J. R. LeMaster and Donald D. Kummings. New York: Garland, 1998. Available on *The Walt Whitman Archive*.
Ignoffo, Matthew F. *What the War Did to Whitman: A Brief Study of the Effects of the Civil War on the Mind of Walt Whitman*. New York: Vantage, 1975.
Jacobus, Mary. "'Distressful Gift': Talking to the Dead." *South Atlantic Quarterly* 106, no. 2 (2007): 393–418.
Jeffery, C. *An Introduction to Plant Taxonomy*. Cambridge: Cambridge University Press, 1982.
Jehlen, Myra. *American Incarnation: The Individual, the Nation, and the Continent*. Cambridge, MA: Harvard University Press, 1986.
Johnson, Lawrence. "Cryptonymic Secretions: On the Kindness of Strangers." In *Re-Reading Derrida: Perspectives on Mourning and Its Hospitalities*, eds. Tony Thwaites and Judith Seaboyer, 117–31. New York: Lexington, 2013.
———. "Tracing Calculation [*Calque Calcul*] between Nicolas Abraham and Jacques Derrida." *Postmodern Culture* 10, no. 3 (2000).
———. *The Wolf Man's Burden*. Ithaca, NY: Cornell University Press, 2001.
Johnston, James Finlay Weir. *The Chemistry of Common Life*. New York: D. Appleton, 1855.
Kaplan, Justin. *Walt Whitman: A Life*. New York: Simon and Schuster, 1980.

Katz, Jonathan. *Love Stories: Sex between Men before Homosexuality.* Chicago: University of Chicago Press, 2001.

Kenny, Maurice. "Whitman's Indifference to Indians." In *The Continuing Presence of Walt Whitman: The Life after the Life,* ed. Robert K. Martin, 28–39. Iowa City: University of Iowa Press, 1992.

Kete, Mary Louise. *Sentimental Collaborations: Mourning and Middle-Class Identity in Nineteenth-Century America.* Durham, NC: Duke University Press, 2000.

Keynes, Richard, ed. *Charles Darwin's Zoology Notes & Specimen Lists from H.M.S. Beagle.* Cambridge: Cambridge University Press, 2000.

Killingsworth, M. Jimmie. *Walt Whitman and the Earth: A Study in Ecopoetics.* Iowa City: University of Iowa Press, 2004.

Kinney, Katherine. "Making Capital: War, Labor, and Whitman in Washington, D.C." In *Breaking Bounds: Whitman and American Cultural Studies*, eds. Betsy Erkkila and Jay Grossman, 174–89. New York: Oxford University Press, 1996.

Kolb, Lawrence C. "Treatment of the Acute Painful Phantom Limb." *Mayo Clinic Proceedings* 25 (1950): 467–71.

Kramer, Lawrence, ed. *Walt Whitman and Modern Music: War, Desire, and the Trials of Nationhood.* New York: Garland, 2000.

Krieg, Joann P. "Charlotte Perkins Gilman and the Whitman Connection." *Walt Whitman Quarterly Review* 1 (1984): 21–25.

———. "Walt Whitman and the Prostitutes." *Literature and Medicine* 14, no. 1 (1995): 36–52.

Kristeva, Julia. *Powers of Horror: An Essay on Abjection*. Trans. Leon S. Roudiez. New York: Columbia University Press, 1982.

Laderman, Gary. *The Sacred Remains: American Attitudes toward Death, 1799–1883.* New Haven, CT: Yale University Press, 1996.

Lamb, D. S. *A History of the United States Army Medical Museum, 1862 to 1917.* Washington, DC: Government Printing Office, 1917.

Launer, John. "Anna O and the 'Talking Cure.'" *QJM* 98, no. 6 (2005): 465–66.

Lavater, John Caspar. *Essays on Physiognomy: Designed to Promote the Knowledge and Love of Mankind.* London: William Tess, 1858.

Lawson, Andrew. "'Spending for Vast Returns': Sex, Class, and Commerce in the First *Leaves of Grass.*" *American Literature* 75, no. 2 (2003): 335–65.

Lee, Han-Mook. "The Inclusive and Elusive Movements of Walt Whitman: Poetic Malleability." *Nineteenth Century Literature in English* 13, no. 2 (2009): 203–24.

LeMaster, J. R., and Donald D. Kummings, eds. *Walt Whitman: An Encyclopedia.* New York: Garland, 1998. *The Walt Whitman Archive.*

Leypoldt, Gunter. "Democracy's 'Lawless Music': The Whitmanian Moment in the U.S. Construction of Representative Literariness." *New Literary History* 38 (2007): 333–52.

Liebig, Justus. *Chemistry in Its Application to Agriculture and Physiology.* Cambridge, MA: John Owen, 1842.

Lincoln, Abraham. *Lincoln: Speeches and Writings 1832–1858.* New York: Penguin Literary Classics, 1989.

Lovejoy, Bess. "The Gory New York City Riot That Shaped American Medicine." *The Smithsonian*, June 17, 2014. Web.

Loving, Jerome. *Walt Whitman: The Song of Himself.* Berkeley: University of California Press, 1999.

Lowenthal, David. "Past Time, Present Place: Landscape and Memory." *Geographical Review* 65, no. 1 (1975): 1–36.

Lucas, Rose. "'Lines of Confusion, Stones of Emptiness': The Place of Mourning in Gwen Harwood's 'Herongate.'" *Southerly* 65, no. 1 (2005): 146–70.

Luciano, Dana. *Arranging Grief: Sacred Time and the Body in Nineteenth-Century America.* New York: New York University Press, 2007.

Lukes, H. N. "Causalgia of the Heart: Phantom Limb Syndrome, Disability and Queer Feeling." *Women & Performance* 19, no. 2 (2009): 227–46.

Machor, James L. *Pastoral Cities: Urban Ideals and the Symbolic Landscape of America.* Madison: University of Wisconsin Press, 1987.

Mack, Stephen John. "A Theory of Organic Democracy." In *A Companion to Walt Whitman*, ed. Donald D. Kummings, 136–50. Oxford: Blackwell, 2009.

MacPhail, Scott. "Lyric Nationalism: Whitman, American Studies, and the New Criticism." *Texas Studies in Literature and Language* 44, no. 2 (2002): 133–60.

Mancuso, Luke. "Civil War." In *A Companion to Walt Whitman*, ed. Donald D. Kummings, 290–311. Oxford: Blackwell, 2006.

Martens, Peter B. "War, Whitman and Osler." *Literature and Medicine* 16, no. 2 (1997): 210–25.

Martin, Robert K., ed. *The Continuing Presence of Walt Whitman: The Life after the Life.* Iowa City: University of Iowa Press, 1992.

Marvin, Carolyn. "The Body of the Text: Literacy's Corporeal Constant." *The Quarterly Journal of Speech* 80, no. 2 (1994): 129–49.

McCabe, James D. *Lights and Shadows of New York Life; or the Sights and Sensations of a Great City.* Philadelphia: National Publishing, 1872.

McConnell, Stuart. "Epilogue: The Geography of Memory." In *The Memory of the Civil War in American Culture*, eds. Alice Fahs and Joan Waugh, 258–66. Chapel Hill: University of North Carolina Press.

McGarry, Molly. *Ghosts of Futures Past: Spiritualism and the Cultural Politics of Nineteenth-Century America.* Berkeley: University of California Press, 2008.

Melzack, Ronald. "Phantom Limbs." *Scientific American* 266 (1992): 120–26.

Miller, Angela. *The Empire of the Eye: Landscape Representation and American Cultural Politics 1825–1875.* Ithaca, NY: Cornell University Press, 1993.

———. "The Panorama, the Cinema, and the Emergence of the Spectacular." *Wide Angle* 18, no. 2 (1996): 34–69.

Miller, J. Hillis. "Absolute Mourning: It Is Jacques You Mourn For." In *Re-Reading Derrida: Perspectives on Mourning and Its Hospitalities,* eds. Tony Thwaites and Jude Seaboyer, 9–23. New York: Lexington, 2013.

Miller, Matt. "Makings of Americans: Whitman and Stein's Poetics of Inclusion." *Arizona Quarterly* 36, no. 3 (2009): 39–59.

Mireur, Hippolyte. *La Syphilis et la prostitution dans leurs rapports avec l'hygiene, la morale et la loi.* Paris: Masson, 1888.

Mitchell, Silas Weir. "The Case of George Dedlow." *Atlantic Monthly* 18, no. 105 (July 1866): 1–11. Reprinted in *The Autobiography of a Quack and The Case of George Dedlow,* 115–49. New York: Century, 1900.

———. *Injuries of Nerves and Their Consequences*. Philadelphia: Lippincott, 1872.
———. "The Medical Department in the Civil War." *Journal of the American Medical Association* 62 (1914): 1445–50.
———. "Phantom Limbs." *Lippincott's Magazine Popular Literature & Science* 8 (1871): 563–69.
Mitchell, Silas Weir, George R. Morehouse, and William W. Keen. *Gunshot Wounds and Other Injuries of the Nerves*. Philadelphia: Lippincott, 1864.
Moon, Michael. *Disseminating Whitman: Revision and Corporeality in Leaves of Grass*. Cambridge, MA: Harvard University Press, 1991.
———. "Memorial Rags." In *Professions of Desire: Lesbian and Gay Studies in Literature*, eds. George Haggerty and Bonnie Zimmerman, 233–40. New York: MLA, 1995.
———. "Solitude, Singularity, Seriality: Whitman Vis-à-Vis Fourier." *English Literary History* 73, no. 2 (2006): 303–23.
Morris, Benjamin Franklin. *Memorial Record of the Nation's Tribute to Abraham Lincoln*. Washington, DC: W. H. & O. H. Morrison, 1896.
Morris, Roy, Jr. *The Better Angel: Walt Whitman in the Civil War*. Oxford: Oxford University Press, 2000.
Mulder T., J. Hochstenbach, P. U. Dijkstra, and J. H. Geertzen. "Born to Adapt, but Not in Your Dreams." *Conscious Cognition* 17, no. 4 (2008): 1266–71.
Murray, Martin G. "Responding Kisses: New Evidence about the Origins of 'Vigil Strange I Kept on the Field One Night.'" *Walt Whitman Quarterly Review* 25 (Spring 2008): 192–97.
———. "Specimen Days." In *A Companion to Walt Whitman*, ed. Donald D. Kummings, 553–66. Oxford: Blackwell, 2009.
———. "Traveling with Wounded: Walt Whitman and Washington's Civil War Hospitals." In *The Walt Whitman Archive*.
Nas, Michael. *Derrida from Now On*. New York: Fordham University Press, 2008.
Nathanson, Tenney. *Whitman's Presence: Body, Voice, and Writing in* Leaves of Grass. New York: New York University Press, 1992.
Neff, John R. *Honoring the Civil War Dead: Commemoration and the Problem of Reconciliation*. Lawrence: University of Kansas Press, 2005.
Nystrom, Kenneth. "The Bioarchaeology of Structural Violence and Dissection in the Nineteenth-Century United States." *American Anthropologist* 116, no. 4 (2014): 1–15.
Oettermann, Stephan. *The Panorama: History of a Mass Medium*. New York: Zone Books, 1997.
Otis, G. A., and D. L. Huntington. *The Medical and Surgical History of the War of the Rebellion*. Vol. 2, part 3. Washington, DC: Government Printing Office, 1883.
Otis Historical Archives. National Museum of Health and Medicine.
Outka, Paul "(De)composing Whitman." *Interdisciplinary Studies in Literature and Environment* 12, no. 1 (2005): 55–56.
Paré, Ambroise. *The Works of That Famous Chirurgeon Ambrose Parey*. Trans. T. Johnson. London: M. Clark, 1649.
Parsons, Amy. "Desire, Forgetting and the Future: Walt Whitman's Civil War." *Arizona Quarterly* 71, no. 3 (2015): 85–109.
Pearce, J. M. S. "Silas Weir Mitchell and the 'Rest Cure.'" *Journal of Neurology, Neurosurgery and Psychiatry* 75 (2004): 381.

Pearce, Roy Harvey. *The Continuity of American Poetry*. Middletown, CT: Wesleyan University Press, 1987.

Peterson, Christopher. *Kindred Specters: Death, Mourning, and American Affinity*. Minneapolis: University of Minnesota Press, 2007.

Petrino, Elizabeth A. "'Silent Eloquence': The Social Codification of Floral Metaphors in the Poems of Frances Sargent Osgood and Emily Dickinson." *Legacy* 15, no. 2 (1998): 139–57.

Philbrick, Nathaniel. "Young Ambition: Charles Wilkes' Antarctic Adventure." *MHQ: Quarterly Journal of Military History* 17 (2004): 84.

Pollak, Vivian R. *The Erotic Whitman*. Berkeley: University of California Press, 2000.

Poole, Robert M. *On Hallowed Ground: The Story of Arlington National Cemetery*. New York: Walker, 2009.

Price, Kenneth M. "Debris, Creative Scatter, and the Challenges of Editing Whitman." In *Walt Whitman: Where the Future Becomes Present*, eds. David Haven Blake and Michael Robertson, 59–80. Iowa City: University of Iowa Press, 2008.

———. *To Walt Whitman, America*. Chapel Hill: University of North Carolina Press, 2004.

———. "Walt Whitman and Civil War Washington." *Leviathan* 16, no. 1 (2014): 121–34.

———. *Whitman and Tradition: The Poet in His Century*. New Haven, CT: Yale University Press, 1990.

Price, Kenneth M., and Jacqueline M. Budell. "Written by Walt Whitman, a Friend." *Prologue* 48, no. 2 (2016).

Ragland, Ellie. *Essays on the Pleasures of Death: From Freud to Lacan*. New York: Routledge, 1995.

Ramachandran, V. S., and Sandra Blakeslee. *Phantoms in the Brain: Probing the Mysteries of the Human Mind*. New York: William Morrow, 1999.

Ramachandran, V. S., and W. Hirstein. "The Perception of Phantom Limbs." *Brain* 121 (1998): 1603–30.

Rand, Nicholas. "Translator's Introduction: Toward a Cryptonymy of Literature." In *The Wolf Man's Magic Word: A Cryptonymy*, by Nicholas Abraham and Maria Torok, trans. Nicholas Rand, li–lxx. Minneapolis: University of Minnesota Press, 1986.

Rasula, Jed. *This Compost: Ecological Imperatives in American Poetry*. Athens: University of Georgia Press, 2002.

Reynolds, David S. *Walt Whitman's America: A Cultural Biography*. New York: Vintage, 1995.

Rhode, Michael. "An Army Museum or a National Collection? Shifting Interests and Fortunes at the National Museum of Health and Medicine." In *Medical Museums: Past, Present, Future*, eds. Samuel J. M. M. Alberti and Elizabeth Hallam, 186–200. London: Royal College of Surgeons of England, 2013.

Richards, Eliza. "Lyric Telegraphy: Women Poets, Spiritualist Poetics, and the 'Phantom Voice' of Poe." *Yale Journal of Criticism* 12, no. 2 (1999): 269–94.

Ricoeur, Paul. *Memory, History, Forgetting*. Trans. K. Blamey and D. Pellauer. Chicago: University of Chicago Press, 2004.

Robb, Graham. *Strangers: Homosexual Love in the Nineteenth Century*. New York: Norton, 2003.

Robertson, Michael. *Worshipping Walt: The Whitman Disciples*. Princeton, NJ: Princeton University Press, 2008.

Roper, Robert. *Now the Drum of War: Walt Whitman and His Brothers in the Civil War*. New York: Walker, 2008.

Rosen, George. "Nostalgia: A 'Forgotten' Psychological Disorder." *Psychological Medicine* 5 (1975): 340–54.

Rosenberg, Charles E. *Cholera Years: The United States in 1832, 1849, and 1866*. Chicago: University of Chicago Press, 1987.

Ruby, Jay. *Secure the Shadow: Death and Photography in America*. Cambridge, MA: MIT Press, 1995.

Russo, Edward J., and Curtis R. Mann. *Oak Ridge Cemetery*. Chicago: Arcadia, 2009.

Sacks, Peter M. *The English Elegy: Studies in the Genre from Spenser to Yeats*. Baltimore: Johns Hopkins University Press, 1985.

Sappol, Michael. *A Traffic of Dead Bodies: Anatomy and Embodied Social Identity in Nineteenth-Century America*. Princeton, NJ: Princeton University Press, 2002.

Sargent, L. M. *Dealings with the Dead, by a Sexton of the Old School*. 2 vols. Boston: Dutton and Wentworth, 1856.

Sawyer, Thomas. Autographed letter signed to Walt Whitman. January 21, 1864. Berg Collection, New York Public Library.

Schantz, Mark S. *Awaiting the Heavenly Country: The Civil War and America's Culture of Death*. Ithaca, NY: Cornell University Press, 2008.

Scheick, William J. "Death and the Afterlife." In *A Companion to Walt Whitman*, ed. Donald D. Kummings, 325–40. Malden, MA: Blackwell, 2006.

Schmidgall, Gary. *Walt Whitman: A Gay Life*. New York: Dutton, 1997.

Schott, G. D. "Revealing the Invisible: The Paradox of Picturing a Phantom Limb." *Brain* 137 (2014): 960–69.

Schroeder-Lein, Glenna R. *Encyclopedia of Civil War Medicine*. New York: M. E. Sharp, 2008.

Schultz, Robert. "Comment: Letter from Virginia." *The Hudson Review* 68, no. 1 (2015): 12–15.

Schuster, David. "Personalizing Illness and Modernity: S. Weir Mitchell, Literary Women, and Neurasthenia, 1870–1914." *Bulletin of the History of Medicine* 79, no. 4 (2005): 695–722.

Sedgwick, Eve Kosofsky. *Epistemology of the Closet*. Berkeley: University of California Press, 1990.

Seltzer, Mark. "Wound Culture: Trauma in the Pathological Public Sphere." *October* 80 (Spring 1997): 3–26.

Sharpe, Tanya Telfair. "Grandison Harris: The Medical College of Georgia's Resurrection Man." In *Bones in the Basement: Postmortem Racism in Nineteenth-Century Medical Training*, eds. Robert Blakely and Judith M. Harrington, 206–26. Washington, DC: Smithsonian Institution, 1997.

Sherwood, Isaac R. *Memories of the War*. Toledo, OH: Chittenden, 1923.

Shively, Charles. *Calamus Lovers: Walt Whitman's Working Class Camerados*. San Francisco: Gay Sunshine, 1987.

Shultz, Suzanne M. *Body Snatching: The Robbing of Graves for the Education of Physicians in Early Nineteenth-Century America*. Jefferson, NC: McFarland, 1992.

Silas Weir Mitchell Papers. College of Physicians of Philadelphia Historical Medical Library.

Simonson, Peter. "A Rhetoric for Polytheistic Democracy: Walt Whitman's 'Poet of Many in One.'" *Philosophy and Rhetoric* 36, no. 4 (2003): 353–75.

Small, Harold Adams, ed. *Abner R. Small, the Road to Richmond: The Civil War Memories of Major Abner R. Small of the Sixteenth Maine Volunteers, Together with the Diary Which He Kept When He Was a Prisoner of War*. Berkeley: University of California Press, 1939.

Sozinsky, Thomas S. "Grave-Robbing and Dissection." *Penn Monthly* 10 (1879): 216–17.

Starobinski, Jean. "The Idea of Nostalgia." *Diogenes* 54 (1966): 81–103.

Stearns, Peter N. "Anatomy and Punishment in Late Eighteenth-Century New York." *Journal of Social History* 22, no. 3 (1989): 507–30.

Strauss, Jonathan. *Human Remains: Medicine, Death and Desire in Nineteenth-Century Paris*. New York: Fordham University Press, 2012.

Sweet, Timothy. *Traces of War: Poetry, Photography and the Crisis of the Union*. Baltimore: Johns Hopkins University Press, 1990.

Sychterz, Jeff. "'Silently Watch[ing] the Dead': The Modern Disillusioned War Poet and the Crisis of Representation in Whitman's *Drum-Taps*." *Discourse* 25, no. 3 (Fall 2003): 9–29.

Taylor, John Traill. *The Veil Lifted: Modern Developments of Spirit Photography*. London: Whitaker, 1894.

Thomas, M. Wynn. "Fratricide and Brotherly Love: Whitman and the Civil War." In *The Cambridge Companion to Walt Whitman*, ed. Ezra Greenspan, 27–44. Cambridge: Cambridge University Press, 1995.

———. *The Lunar Light of Whitman's Poetry*. Cambridge, MA: Harvard University Press, 1987.

Thomas Biggs Harned Collection of the Papers of Walt Whitman. Library of Congress.

Trachtenberg, Alan. "Albums of War: On Reading Civil War Photographs." *Representations* 9 (1985): 1–32.

———. "Lincoln's Smile: Ambiguities of the Face in Photography." In *Lincoln's Smile and Other Enigmas*, 69–86. New York: Hill and Wang, 2007.

Traubel, Horace, ed. *With Walt Whitman in Camden*. 9 vols. Vol. 1, Boston: Small Maynard, 1906; Vol. 2, New York: D. Appleton, 1908; Vol. 3, New York: Mitchell Kinnerley, 1914; Vols. 4–7, Carbondale: Southern Illinois University Press, 1959, 1964, 1986, and 1992, respectively; Vols. 8–9, Oregon House, CA: W. L. Bentley, 1996.

Travis, Jennifer. *Wounded Hearts: Masculinity, Law and Literature in American Culture*. Durham, NC: University of North Carolina Press, 2005.

Turpin, Zachary. "Introduction to Walt Whitman's 'Manly Health and Training.'" *Walt Whitman Quarterly Review* 33 (2016): 147–83.

Tward, A. D., and H. A. Patterson. "From Grave Robbing to Gifting: Cadaver Supply in the United States." *Journal of the American Medical Association* 287, no. 9 (2002): 1183–85.

"United States South Seas Exploring Expedition (aka the Wilkes Expedition), 1838–1842." Harvard University Libraries Open Collections Program. http://ocp.hul.harvard.edu/expeditions/wilkes.html.

Van der Kolk, Bessel A., Alexander C. McFarlane, and Lars Weisaeth. *Traumatic Stress: The Effects of Overwhelming Experience on Mind, Body, and Society*. New York: Guilford, 1996.

Waite, Frederick C. "Grave Robbing in New England." *Bulletin of the Medical Library Association* 33, no. 3 (1945): 272–94.

"Walt Whitman." *New York Mirror.* April 1, 1892.

Warner, Michael. "Civil War Religion and Whitman's *Drum-Taps.*" *In Walt Whitman: Where the Future Becomes Present,* eds. David Haven Blake and Michael Robertson, 81–91. Iowa City: University of Iowa Press, 2008.

Warren, James Perrin. "Reading Whitman's Postwar Poetry." In *The Cambridge Companion to Walt Whitman*, ed. Ezra Greenspan, 45–65. Cambridge: Cambridge University Press, 1995.

Warren, Leonard. *Joseph Leidy: The Last Man Who Knew Everything.* New Haven, CT: Yale University Press, 1998.

Watkin, William. *On Mourning: Theories of Loss in Modern Literature*. Edinburgh: Edinburgh University Press, 2004.

Werbel, Amy Beth. *Thomas Eakins: Art, Medicine and Sexuality in Nineteenth-Century Philadelphia.* New Haven, CT: Yale University Press, 2007.

Whitman, Walt. *An American Primer.* Ed. Horace Traubel. Boston: Small, Maynard, 1904.

———. Autographed letter signed to Lewis K. Brown. Brooklyn, July 11, 1864. Berg Collection. New York Public Library.

———. "A Carol of Harvest, for 1867." *Galaxy* 4 (September 1867): 605–9.

———. *The Collected Writings of Walt Whitman: The Early Poems and Fiction.* Ed. Thomas L. Brasher. New York: New York University Press, 1963.

———. *The Collected Writings of Walt Whitman: The Journalism 1834–1846*. Ed. Henry Bergman, Douglas A. Noverr, and Edward J. Recchia. New York: Peter Lang, 1998.

———. *Complete Poetry and Collected Prose*. Ed. Justin Kaplan. New York: Literary Classics of the United States, 1982.

———. *The Correspondence of Walt Whitman*. Ed. Edwin Haviland Miller. 6 vols. New York: New York University Press, 1961.

———. Draft letter to Thomas P. Sawyer. April 26, 1863. Berg Collection. New York Public Library.

———. Draft letter to Thomas P. Sawyer. May 27, 1863. Berg Collection. New York Public Library.

———. Draft letter to Thomas P. Sawyer. July 1863. Berg Collection. New York Public Library.

———. *Leaves of Grass*. Brooklyn, NY, 1855. Available on *The Walt Whitman Archive.*

———. *Leaves of Grass*. Brooklyn, NY, 1856. Available on *The Walt Whitman Archive.*

———. *Leaves of Grass*. Boston: Thayer and Eldridge, 1860. Available on *The Walt Whitman Archive.*

———. *Leaves of Grass*. New York: William E. Chapin, 1867. Available on *The Walt Whitman Archive.*

———. *Leaves of Grass*. Boston: James R. Osgood, 1881–82. Available on *The Walt Whitman Archive.*

———. *Memoranda during the War*. Camden, NJ, 1875–76. Available on *The Walt Whitman Archive.*
———. "My visits and distributions." Holograph fragment. 1863. Berg Collection. New York Public Library.
———. *The Norton Critical Edition of Leaves of Grass*. 1855, 1891–92. Ed. Michael Moon. New York: Norton, 2002.
———. *Notebooks and Unpublished Prose Manuscripts*. Ed. Edward F. Grier. 6 vols. New York: New York University Press, 1984.
———. *November Boughs.* 1888. In *Walt Whitman: Complete Poetry and Collected Prose,* 1135–1244. New York: Library of America, 1982.
———. *Prose Works: Collected and Other Prose*. Ed. Floyd Stovall. Vol. 2. New York: New York University Press, 1964.
———. *Specimen Days and Collect*. Boston: D. McKay, 1883. Available on *The Walt Whitman Archive.*
———. *The Uncollected Poetry and Prose of Walt Whitman.* Ed. Emory Holloway. 2 vols. Garden City, NY: Doubleday, 1921.
"Whitman's Farewell." *The New York Tribune*. August 16, 1891.
Williams, William Carlos. "An Essay on *Leaves of Grass*." In *Leaves of Grass: One Hundred Years After*, ed. Milton Hindus. Stanford, CA: Stanford University Press, 1955.
Wilson, Ivy G. "Organic Compacts, Form, and the Cultural Logic of Cohesion; or, Whitman Re-Bound." *ESQ: A Journal of the American Renaissance* 54, no. 1 (2008): 199–215.
———, ed. *Whitman Noir: Black America and the Good Gray Poet*. Iowa City: University of Iowa Press, 2014.
Wrobel, Arthur. "Whitman and the Phrenologists: The Divine Body and the Sensuous Soul." *PMLA* 89, no. 1 (1974): 17–23.
Wurzer, Wilhelm S., ed. "Epilogue: The Paradox of Philosophy's Gaze." In *Panorama: Philosophies of the Visible*, ed. Wurzer, 201–10. New York: Continuum, 2002.
Yalom, Marilyn. *The American Resting Place: Four Hundred Years of History through Our Cemeteries and Burial Grounds.* New York: Houghton Mifflin, 2008.

INDEX

THE IOWA WHITMAN SERIES

Conserving Walt Whitman's Fame: Selections from Horace Traubel's "Conservator," 1890–1919, edited by Gary Schmidgall

Democratic Vistas: The Original Edition in Facsimile, by Walt Whitman, edited by Ed Folsom

Intimate with Walt: Selections from Whitman's Conversations with Horace Traubel, 1888–1892, edited by Gary Schmidgall

Leaves of Grass, 1860: The 150th Anniversary Edition, by Walt Whitman, edited by Jason Stacy

Life and Adventures of Jack Engle: An Auto-Biography, by Walt Whitman, with an introduction by Zachary Turpin

A *Place for Humility: Whitman, Dickinson, and the Natural World*, by Christine Gerhardt

The Pragmatic Whitman: Reimagining American Democracy, by Stephen John Mack

"Song of Myself": With a Complete Commentary, by Walt Whitman and introduction and commentary by Ed Folsom and Christopher Merrill

Supplement *to "Walt Whitman: A Descriptive Bibliography,"* by Joel Myerson

Transatlantic Connections: Whitman U.S., Whitman U.K., by M. Wynn Thomas

Visiting Walt: Poems Inspired by the Life and Work of Walt Whitman, edited by Sheila Coghill and Thom Tammaro

Walt Whitman and the Class Struggle, by Andrew Lawson

Walt Whitman and the Earth: A Study in Ecopoetics, by M. Jimmie Killingsworth

Walt Whitman: The Correspondence, Volume VII, edited by Ted Genoways

Walt Whitman, Where the Future Becomes Present, edited by David Haven Blake and Michael Robertson

Walt Whitman's Reconstruction: Poetry and Publishing between Memory and History, by Martin T. Buinicki

Walt Whitman's Selected Journalism, edited by Douglas A. Noverr and Jason Stacy

Walt Whitman's Songs of Male Intimacy and Love: "Live Oak, with Moss" and "Calamus," edited by Betsy Erkkila

Whitman among the Bohemians, edited by Joanna Levin and Edward Whitley

Whitman East and West: New Contexts for Reading Walt Whitman, edited by Ed Folsom

Whitman Noir: Black America and the Good Gray Poet, edited by Ivy G. Wilson

Whitman's Drift: Imagining Literary Distribution, by Matt Cohen